Metabolomics - New Insights into Biology and Medicine

Edited by Wael N. Hozzein

Published in London, United Kingdom

IntechOpen

Supporting open minds since 2005

Metabolomics - New Insights into Biology and Medicine
http://dx.doi.org/10.5772/intechopen.73760
Edited by Wael N. Hozzein

Part of IntechOpen Book Series: Biochemistry, Volume 16
Book Series Editor: Miroslav Blumenberg

Contributors
Rosanna Floris, Carmen Rizzo, Angela Lo Giudice, Elizabeth Lusczek, Natalia Beloborodova, Andrey Yu. Olenin, A.V. Grechko, Mercedes G. López, Alicia Huazano-Garcia, Juan Vázquez-Martínez, Claudio Horacio Morales, Yibin Feng, Wei Guo, Hor Yue Tan, Ning Wang

First published in London, United Kingdom, 2020 by IntechOpen
IntechOpen is the global imprint of INTECHOPEN LIMITED, registered in England and Wales, registration number: 11086078, 7th floor, 10 Lower Thames Street, London, EC3R 6AF, United Kingdom
Printed in Croatia

British Library Cataloguing-in-Publication Data
A catalogue record for this book is available from the British Library

Additional hard and PDF copies can be obtained from orders@intechopen.com

Metabolomics - New Insights into Biology and Medicine
Edited by Wael N. Hozzein
p. cm.
Print ISBN 978-1-78985-127-4
Online ISBN 978-1-78985-128-1
eBook (PDF) ISBN 978-1-83962-298-4
ISSN 2632-0983

For EU product safety concerns:
IN TECH d.o.o., Prolaz Marije Krucifikse Kozulić 3, 51000 Rijeka, Croatia, info@intechopen.com or visit our website at intechopen.com.

IntechOpen Book Series
Biochemistry
Volume 16

Prof. Wael N. Hozzein is a professor of microbiology at the Faculty of Science, Beni-Suef University, Egypt. He received his PhD from Cairo University, Egypt, in 2003, and then worked as a visiting scientist at Newcastle University, UK, and Michigan State University, USA. Recently, he worked as the chair professor of the Bioproducts Research Chair at King Saud University, Saudi Arabia. He has vast experience in the field of bacterial taxonomy with research interests in microbial biodiversity and applications. Prof. Hozzein is the author of more than 140 publications and a guest editor and reviewer for several international journals. Additionally, he has been involved in many academic activities and educational reform projects and initiatives. He has been the principal investigator for several funded grants and has also received several awards, for example the State Encouragement Prize in Biological Sciences for 2015, Egypt.

Editor of Volume 16:
Wael N. Hozzein
Botany and Microbiology Department, Faculty of Science, Beni-Suef University, Beni-Suef, Egypt

Book Series Editor: Miroslav Blumenberg
NYU Langone Medical Center, New York, USA

Scope of the Series

Biochemistry, the study of chemical transformations occurring within living organisms, impacts all of life sciences, from molecular crystallography and genetics, to ecology, medicine and population biology. Biochemistry studies macromolecules - proteins, nucleic acids, carbohydrates and lipids –their building blocks, structures, functions and interactions. Much of biochemistry is devoted to enzymes, proteins that catalyze chemical reactions, enzyme structures, mechanisms of action and their roles within cells. Biochemistry also studies small signaling molecules, coenzymes, inhibitors, vitamins and hormones, which play roles in the life process. Biochemical experimentation, besides coopting the methods of classical chemistry, e.g., chromatography, adopted new techniques, e.g., X-ray diffraction, electron microscopy, NMR, radioisotopes, and developed sophisticated microbial genetic tools, e.g., auxotroph mutants and their revertants, fermentation etc. More recently, biochemistry embraced the 'big data' omics systems.

Initial biochemical studies have been exclusively analytic: dissecting, purifying and examining individual components of a biological system; in exemplary words of Efraim Racker, (1913 - 1991) "Don't waste clean thinking on dirty enzymes." Today however, biochemistry is becoming more agglomerative and comprehensive, setting out to integrate and describe fully a particular biological system. The "big data" metabolomics can define the complement of small molecules, e.g., in a soil or biofilm sample; proteomics can distinguish all the proteins comprising e.g., serum; metagenomics can identify all the genes in a complex environment e.g., bovine rumen. This Biochemistry Series will address both the current research on biomolecules, and the emerging trends with great promise.

Contents

Preface

Metabolomics is a rapidly evolving branch of the "omics technologies," which involves quantitative and qualitative metabolite assessments. Metabolomics comprises analytical techniques that can be used for metabolite profiling, fingerprinting, and quantitation of targeted metabolites, and the study of the biochemical fate of individual metabolites. Metabolomics data analysis and databases are major areas of research based on the availability of high-quality reference metabolite standards and data analysis tools. Discovery of new drugs from different natural resources is a challenging research of interest in the field of metabolomics. In this regard, the discovery of new antiviral drugs and vaccines is an emerging demand now after the effects the world is facing due to the pandemic of COVID-19.

Metabolomics offers diverse applications in different fields such as biological, medical, and environmental sciences. It could be used in the detection of disease-related biomarkers and in understanding the metabolic changes in relation to diseases. It also helps in the assessment of contaminants and pollutants among many other environmental applications. The integration of metabolomics with genomics and proteomics can enable us to understand the whole biochemical pathway from the genetic order "genotype" to its expression "phenotype," which will lead in the future to amazing new insights into biology and medicine.

This book is mainly for researchers interested in the new developments and applications of metabolomics. It is also important for physicians using metabolomic approaches in the diagnosis of diseases or treatment and for postgraduate students starting their research projects on metabolomics. The book is divided into two sections as indicated from its title, namely: new insights into biology and new insights into medicine.

The first chapter discusses the potential of marine microorganisms for the production of biosurfactants, which are important functional molecules with several applications. The chapter gives a general overview of the recent advances in biosurfactants of marine origin, their production, chemical identification, interesting biological properties, and potential biotechnological applications.

The second chapter is a research study conducted to identify the enriched or depleted metabolites in the feces of overweight mice after dietary treatment as biomarkers of interactions between diet and health. The authors found that the microbial metabolites, coming from the microbial fermentation of agavins, induced a beneficial effect on the health of overweight mice. These findings open exciting opportunities to explore new biomarkers with applicability to the prevention of overweight-associated metabolic syndromes and treatment of overweight people.

On the other hand, the next three chapters follow applications of the metabolic approach in the medical field. Chapter 3 discusses the mutually beneficial integration of the metabolome/microbiome in the body of healthy people and focuses on the effects of microbiota dysfunction as the cause of pathology. This chapter explains the effect of certain microbial metabolites on the work of key enzymes, which becomes more important in patients at risk of developing some disorders or in the development of some life-threatening conditions and serious diseases. Interestingly, the authors of this chapter call the microbiota an "invisible organ," emphasizing its functional significance.

In Chapter 4, the author investigates the applicability of serum metabolomics as a powerful tool in the diagnosis of trauma and other critical conditions, such as the respiratory failure, pancreatitis, and combat trauma of patients hospitalized in the intensive care unit (ICU). The meta-analysis results of the serum metabolomes revealed how the metabolomes differ with each condition. The results revealed also how mass spectrometry-based metabolomics could be used for predictive monitoring of critical illness conditions. The study highlights that metabolomics could lead to faster and more appropriately individualized patient access to the ICU, which will improve patient care and outcomes.

Lastly, Chapter 5 systematically reviews recent studies on metabolism-targeting anticancer therapies based on metabolomics in terms of glucose, lipid, amino acid, nucleotide, and other altered metabolisms, with special emphasis on the potential of metabolic treatment with pure compounds, herbal extracts, and formulations from Chinese medicine. The trends of future development of metabolism-targeting anticancer therapies are also discussed. The systematic review in this chapter of recent metabolomic studies targeting Chinese medicine draws attention to this traditional medicine as a promising candidate for more effective and safer treatment of human cancers clinically.

I would like to thank all the authors of the chapters of this book for sharing their expertise and for their quality of work. On behalf of all authors, I would like to warmly thank Ms. Romina Skomersic and Mr. Gordan Tot, the author service managers, for their sincere help in the publication of this book. Our thanks are also extended to other staff of IntechOpen for their support.

Finally, the editing of this book was of special interest to me and I wish that readers will also find it stimulating. Overall, this is a well-written book, containing some very interesting research avenues and I hope it will contribute positively to the academic and research communities.

Wael N. Hozzein
Botany and Microbiology Department,
Faculty of Science,
Beni-Suef University,
Beni-Suef, Egypt

Section 1

Metabolomics: New Insights into Biology

Chapter 1

Biosurfactants from Marine Microorganisms

Rosanna Floris, Carmen Rizzo and Angelina Lo Giudice

Abstract

The marine biosphere represents a yet underexploited natural source of bioactive compounds, mainly of microbial origin. Among them, biosurfactants (BSs) are functional molecules, which are attracting a great interest due to their biocompatibility, versatility, and applications in several biotechnological fields. BSs are surface active amphipathic compounds, containing both a hydrophilic and a hydrophobic moiety, which are grouped in low (glycolipids and lipopeptides) or high molecular weight (polymeric complexes) compounds. A number of environmental factors such as pH, salinity, temperature, and nutrient availability can affect microbial BS production. Marine microorganisms with different phylogenetic affiliations and isolated from several marine habitats (e.g., seawater, sediments, and higher organisms) worldwide (spanning from the Mediterranean Sea to Antarctica) have been reported as surfactant producers. However, most of the marine microbial world remains still unexplored. The present chapter aims at giving a general overview on the recent advances about BSs of marine origin, in order to enhance the knowledge inherent their production, chemical characterization and identification, interesting biological properties, and potential biotechnological applications.

Keywords: bacteria, surfactants, aquatic environments, biodegradability, natural tensioactive, biotechnology

1. Introduction

In aquatic ecosystems, marine microorganisms have developed unique metabolic and physiological capabilities to adapt to diverse habitats which cover a wide range of thermal, pressure, salinity, pH, and nutrient conditions [1]. In this way, the symbiotic association and the interaction of biological systems with bacteria have induced the production of a variety of bioactive compounds such as biosurfactants (BSs), antibiotics, enzymes, vitamins, [2] and more than 10,000 metabolites with a broad spectrum of biological activities that have been isolated from marine microbes. By evolution, bacteria have adapted themselves to feeding on water-immiscible materials by producing and using surface-active products that help them in aqueous phase to adsorb, emulsify, wet, and disperse or solubilize water-immiscible materials [3]. In this context, surface active metabolites as BSs have gained much attention because of their biodegradability, low toxicity, and ability to be produced from renewable and cheaper substrates, thus getting an important ecological role due to their structural, functional diversity, and the potential multidisciplinary applications in industrial and environmental fields. These features in addition to their stability at extremes of temperatures, pH, and salinity make them commercially optimal alternatives to their chemically synthesized counterparts. For this reason, different authors have tested BSs of marine origin for environmental applications such as bioremediation processes, dispersion of oil spills, and enhancement of oil recovery. Furthermore, some BSs play an essential role for the survival of microbial producers against other competing or dangerous microorganisms by acting as biocide agents [4], and some others have shown the ability to stimulate plant and animal defense responses against microbes [5]. By considering that, these aspects have been explored especially for terrestrial bacteria, in the light of the above said characteristics, BSs from marine bacteria have been getting attention more and more as new suitable alternatives to chemical surfactants of petroleum origin in the food, cosmetic, health care industries [6], synthetic medicines and can be safe and effective therapeutic agents in medicine.

The purpose of this chapter is to provide a comprehensive overview on the recent advances about different types of marine BSs, to highlight the recent studies on the new sources of production, and to focus on the state of art of the screening methodologies for the identification, characterization, and potential biotechnological applications.

2. Biosurfactants

BSs are secondary metabolism bacterial products which exhibit surface and emulsifying activities thanks to the hydrophobic part of the molecule and the hydrophilic water soluble end. They are produced extracellularly or as a part of the cell membrane by a variety of yeasts, bacteria, and filamentous fungi from various substrates as sugars, oils, alkanes, and wastes [7]. The name *surfactant* derives from their surface chemical action, as they tend to interact at the boundary level between two phases in a heterogeneous system by forming a film which can change the properties (wettability and surface energy) of the original surface. They are mainly classified in BSs acting by reducing surface tension at the air-water interface (biosurfactants), and BSs that reduce the interfacial tensions between immiscible liquids or at the solid-liquid interface (bioemulsifiers) [8]. The investigations on BS production from microorganisms are more and more frequent and are subdivided into three main steps: (1) the potential BS producer isolation; (2) the screening for BS production; and (3) the extraction and purification step, sometimes improved with the chemical characterization of molecules. According to Ref. [6], the producer isolation and detection are crucial phases in the research for new BSs, which have to be strongly related to the aim of the investigation. While the terrestrial sources have been extensively explored, the marine environments have been focused as potential optimal source only in recent decades. It has been reported that BS-producing bacteria are widely distributed in both contaminated and undisturbed water or soil [9, 10]. Indeed, the most exploited source of microbial BS producers is represented by sediment and water samples, with different types and levels of contamination [11–13]. As a matter of fact, it is retained that microbial degraders of insoluble substrates, that is, hydrocarbons or oils, are stimulated to produce secondary metabolites that are able to enhance the cellular uptake and degradation of such compounds. In some cases, enrichment cultures have been performed with natural samples in order to favor the isolation of potential producers [14, 15]. Despite the application of this principle has gained optimal results in the discovery of new marine producers and BSs [14, 16], some authors assessed that the BS production is not strictly linked to the hydrocarbon uptake [17, 18]. Only recently, innovative marine sources of isolation has been proposed, and researchers have focused their studies on biological matrices, that is, filter-feeding organisms as host of microbial communities specialized in the production of secondary metabolites with functional roles. As reviewed in [6], BS producers with optimal potentialities have been isolated from polychaetes [19, 20], sponges [21–25], sea pens [26], cnidarians [27, 28], and fish [29]. A common characteristic of BSs is to relax or decrease surface tension and this increases solubility so that BSs may interact with the interfaces and affect the adhesion and the detachment of bacteria [4]. All these properties confer to the BS antibacterial, antifungal, and antiviral activities [30], in addition to the pollutants removal potential. For these reasons, particular growing interest at global level is centered into new and unexplored sources of BSs as marine and deep-sea environments represent.

2.1 BS classification/types, census of BS marine microbial producers

Surfactants are generally classified according to the nature of the charge on polar moiety: anionic (negatively charged), nonionic (polymerization products), cationic (positively charged), and amphoteric (both negatively and positively charged). They can be grouped into two categories: (a) low molecular mass molecules (rhamnolipids, sophorolipids, trehalose lipids, lipopeptides phospholipids, fatty acids, and neutral lipids) [31], which lower the surface and interfacial tensions and (b) high molecular polymer mass (high molecular weight polysaccharide, polysaccharide-protein complexes, lipopolysaccharides, and lipoproteins called emulsans), which bind tightly to surfaces [32]. Various microbial species produce different BSs [33]. The most known ones are glycolipids, a class of molecules made of mono-, di-, tri- and tetra-saccharides which include glucose, mannose, galactose, glucuronic acid, rhamnose, and galactose sulfate in combination with long-chain aliphatic acids or hydroxyaliphatic acids. Among glycolipids, rhamnolipids, and trehalolipids, sophorolipids are the most studied disaccharides. Rhamnolipids are known to be produced by the *Pseudomonas* genus [31, 34], while trehalolipids are generally produced by many members of the genera *Mycobacterium*, *Nocardia* and *Corynebacterium*, *Rhodococcus erythropolis*, and *Arthrobacter* sp. [35]. Finally, sophorolipids represent a group of BSs by *Candida* spp. [31]. Another class of surfactant compounds which received considerable attention is the fatty acids such as the phospholipids derived from alkane substrates. Indeed, they represent the major components of microbial membranes which are mainly produced by hydrocarbon-degrading bacteria as *Acinetobacter* sp. *Rhodococcus erythropolis*, *Thiobacillus thiooxidans, Capnocytophaga* sp., *Penicillium spiculisporum,* and *Corynebacterium* sp., or yeasts [32]. Moreover, a number of microorganisms (e.g., *Candida lipolytica* and *Saccharomyces cerevisiae*) with different taxonomic affiliations produce exocellular polymeric surfactants called bioemulsans, composed of polysaccharides, protein, lipopolysaccharides, and lipoproteins, among which the bioemulsans produced by different species of *Acinetobacter* are the most studied [36]. Different authors described microbial BSs from marine sources (sediments, corals, sponges, sea, and hot water springs) which include several lipopeptide antibiotics with potent surface-active properties [22, 37–41]. Some authors [22] detected carbohydrate, protein, and lipid contents (20 μg/0.1 ml, 35 μg/0.1 ml, and 573 μg/0.1 ml, respectively) in the analyzed BS, suggesting a chemical structure typical of a lipopeptidic compound for the sponge-associated marine actinomycetes *Nocardiopsis alba* MSA10. Further chemical analyses [Fourier-transformed infrared spectrophotometer analysis (FT-IR), nuclear magnetic resonance (NMR), and gas chromatography mass spectrometry (GC/MS)] were carried out by other scientists [21, 23, 28], and lipopeptidic BSs were detected for *Bacillus* spp. strains, in line with previous data on the neighboring cluster *Bacillus* as prominent lipopeptide producers. Interestingly, the lipopeptide identified in [23] was firstly demonstrated from *Brevibacterium* stain MSA13 (so-called *brevifactin*) and showed a different structure respect to that observed for other lipopeptides. Indeed, it was composed of a hydrophobic moiety of octadecanoic acid methyl ester and a peptide part contained a short sequence of four amino acids including pro-leu-gly-gly (**Figure 1**). Similarly, the possible production of novel BSs by *P. rettgeri*, *Psychrobacter* sp., and *B. anthracis* isolates was suggested by the analysis of FTIR spectra [28].

In the marine context, the group of glycolipids has been widely studied because they are produced by a broad spectrum of bacteria isolated from various marine matrices, both animals (Annelida, sea pen *Pteroeides*, and fish gut) and contaminated soils (Arctic and Antarctic sediments) [14, 19, 20, 24, 26, 29, 42]. A glycolipopeptide BS produced by the coral-associated *Bacillus* sp. E34 was identified by means of FT-IR analysis [27].

A list of various studied BS types with the respective microbial producers and the source of isolation is shown in **Table 1**.

Finally, other studies drew their attention on a group of heterogeneous high molecular weight bioactive compounds not strictly defined as biosurfactants but endowed with interfacial properties, called exopolysaccharides (EPSs), often originated from marine both prokaryotes and eukaryotes (cyanobacteria and microalgae) and extreme environments [43].

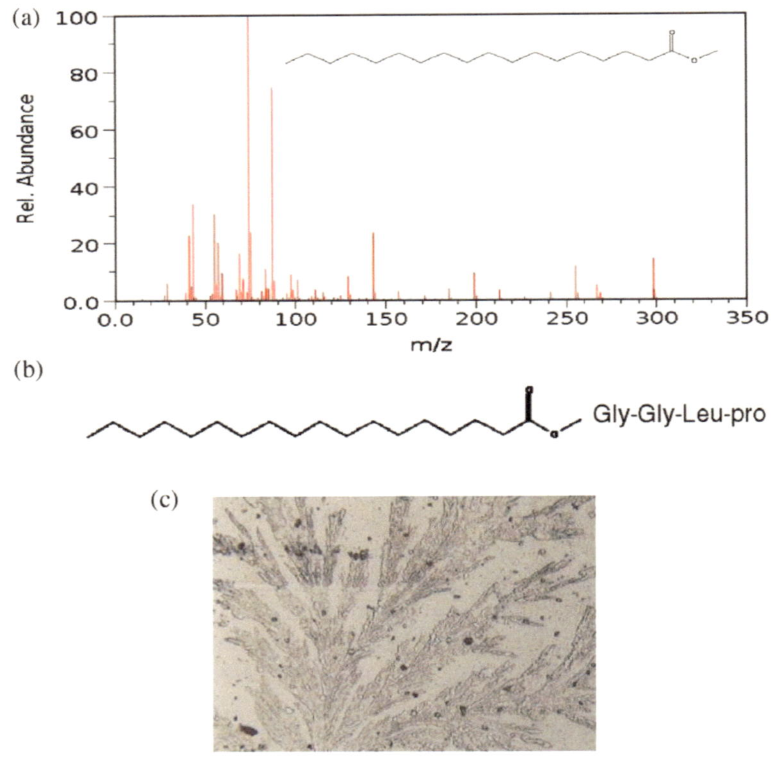

Figure 1.
MS spectra of octadecanoic acid methyl ester (a), probable structure of octadecanoic acid methyl ester with peptide moiety (b), crystalline appearance of the recovered crystals of lipopeptide (MSA13) examined under a light microscope at 40× [23].

Origin	BS	Bacterial Affiliations	Reference
Natural samples			
Seawater, sediment enrichments		*Alcanivorax borkumensis*, *Paracoccus marcusii*	[65]
Sand/soil samples and seawater	ND	*Catenovulum* sp., *Cobetia* sp., *Glaciecola* sp., *Serratia* sp., *Marinomonas* sp., *Pseudomonas* sp., *Pseudoalteromonas* sp., *Psychromonas* sp., *Rhodococcus* sp.,	[11]
Oil-spilled seawater samples	ND	*Acinetobacter* sp., *Arthrobacter* sp., *Bacillus* sp, *Gluconobacter* sp., *Pseudomonas* sp.	[12]
Seawater	Glycoprotein	*Pseudoalteromonas* sp.	[42]
Black sea, Red sea, hot water springs	ND	*Bacillus* spp.	[40]
Shoreline marine sediment enrichments	Glycolipid	*Pseudomonas* spp., *Pseudoalteromonas* spp., *Idiomarina sp.*, *Rhodococcus* spp.	[14]
Seawater sediment, shells samples	ND	*Acinetobacter* spp., *Bacillus* spp, *Pseudomonas* sp.	[30]
Seawater samples	ND	*Pseudomonas aeruginosa*, *Aeromonas hydrophila*	[13]
Seawater samples	Lipopeptide	*Halobacterium salinarum*	[50]
Seawater enrichments	ND	*Bacillus cereus*, *B. megaterium*, *B. subtilis*, *Branhamella catarrhalis*, *Citrobacter intermedius*, *Corynebacterium kutscheri*, *C. xerosis*, *Enterobacter aerogenes*, *Escherichia coli*, *Klebsiella ozaenae*, *Lactobacillus casei*, *L. delbrueckii*, *Proteus inconstans*, *Pseudomonas aeruginosa*, *P. fluorescens*, *P. diminuta*, *P. mallei*, *Staphylococcus aureus*	[15]
Biological Matrices			
Callyspongia diffusa	Lipopeptide	*Alcaligenes* sp., *Bacillus* sp., *Halomonas* sp.	[21]
Sparus aurata	Glycolipid/Rhamnolipid	*Pseudomonas* spp., *Acinetobacter* sp., *Sphingomonas* spp., *Aeromonas* sp.	[29]
Pteroeides spinosum	Glycolipid	*Brevibacterium* sp., *Vibrio* sp.	[26]
Callyspongia diffusa	Lipopeptide	*Nocardiopsis alba*	[22]
Dendrilla nigra	New Lipopetide (Brevifactin)	*Brevibacterium aureum*	[23]
Dendrilla nigra	Glycolipid	*Brachybacterium paraconglomeratum*	[24]
Sarchophyton glaucum	Glycolipopeptide	*Bacillus* sp. E34	[27]
Acropora digitifera	Lipopeptide/New BS	*Providencia rettgeri*, *Psychrobacter* sp., *Bacillus flexus*, *Bacillus anthracis*, *Psychrobacter* sp., *Bacillus pumilus*	[28]
Megalomma claparedei, *Sabella spallanzanii Branchiomma luctuosum* enrichment cultures	ND	*Joostella* spp., *Cellulophaga* spp., *Thalassospira* spp., *Pseudovibrio* spp., *Pseudomonas* spp., *Alcanivorax* spp., *Cellulophaga* spp., *Cobetia* sp., *Cohaesibacter* spp., *Idiomarina* spp., *Marinobacter* sp.	[19]
Megalomma claparedei, *Sabella spallanzanii Branchiomma luctuosum*		*Psychrobacter* sp., *Pseudoalteromonas* spp., *Vibrio* sp., *Maribacter* sp., *Cellulophaga* spp., *Tenacibaculum* spp., *Citricoccus* spp., *Staphylococcus* sp.	[20]

ND: not determined.

Table 1.
Overview of various studied BS types with the respective marine producers and the source of isolation.

3. Natural roles and biotechnological applications

Biosurfactants have various functions which are often unique for the physiology and ecology of their microbial producers. As stated above, one of the most interesting roles from the environmental point of view is represented by the different strategies adopted by microorganisms to enhance the bioavailability and the access to hydrophobic compounds as carbon source [44]. A mechanism proposed by which hydrocarbons became incorporated within the hydrophobic core of the BS micelles is shown in **Figure 2**. This process studied with alkanes as model substrates and referred to as "micelle solubilization" [45], favors their dispersion into the aqueous phase and their bioavailability for uptake by cells.

The potential application of BSs in hydrocarbon bioremediation has been investigated for marine microorganisms from different origins and allowed to obtain interesting results. In this case, the use of contaminated samples for isolation of potential BS producers is extremely encouraged. Strains affiliated to *Rhodococcus* spp. were reported as capable of reduce surface tension in the presence of oily substrates, and the extracted BSs have been proved optimal enhancer

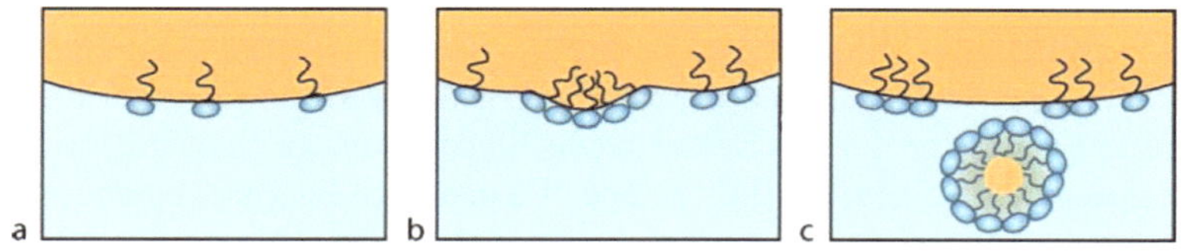

Figure 2.
Mechanism of hydrocarbon solubilization so called "micelle solubilization" within biosurfactant micelles: biosurfactants at the interface between the aqueous and hydrocarbon phases [45].

for n-hexadecane biodegradation at 13°C and of tetradecane [11, 14]. The studies described in [26] evaluated the BS production in the presence of hydrocarbonic substrates to test the potential of *Brevibacterium* and *Vibrio* spp. strains in the field of bioremediation and reported the diesel oil as better utilized carbon source. The hydrocarbon remediation aspect was also deeply studied by investigating *Joostella* sp. A8, *Alcanivorax* sp. A53, and *Pseudomonas* sp. A6 for BS production and diesel oil degradation in pure culture and co-culture conditions [46]. Interestingly, the biodegradation rates and the efficiency increased in co-culture of 99.4 and 99.2%, respectively. Furthermore, in Ref. [27], a glycolipidic biosurfactant produced by the coral-associated *Bacillus* with a removing capacity of about 45% was reported as optimal candidate for oil removal. The bioremediation potential of bacterial BSs has been investigated also in terms of chelating activity toward heavy metals, despite in this context, the literature is scarce. In fact, several bacteria have been reported as able to produce BSs in the presence of heavy metals such as Cd, Cu, and Zn, as *Joostella* sp. A8 and *Alcanivorax* sp. A53 [47, 48]. Despite, the heavy metal removal is an interesting topic, it remains still largely unexplored [47–49] for marine bacteria and need to be improved especially for metal chelation in aqueous systems. BS-producing microorganisms have developed other important physiological functions as response to needs and life style, such as antimicrobial activity (mainly lipopeptide and glycolipid surfactants), biofilm formation or processes of motility and colonization of surfaces. Immunomodulation and enzyme inhibition, have been detected for several BSs from marine environments or not. The antimicrobial activity of BS has been studied *in vivo* and *in vitro* and a broad spectrum of this activity against Gram-positive, Gram-negative, fungi, viruses, algae, etc., so as different modes of action were detected [5]. Both lipopeptides and glycolipids of marine origin have been proved as active against several bacterial pathogens. Indeed, the lipopeptide produced by *N. alba* [22] exhibited antibacterial activity against *E. faecalis*, *B. subtilis*, and the pathogenic yeast *C. albicans*. Similarly, the sponge-associated *Brevibacterium aureum* MSA13 and *Brachybacterium paraconglomeratum* MSA21 were proved as BS producers with a wide antibacterial activity toward several pathogens such as *B. subtilis*, *C. albicans E. coli*, *E. faecalis*, *K. pneumonia, M. luteus*, *P. aeruginosa*, *P. mirabilis*, *Streptococcus* sp., *S. aureus,* and *S. epidermidis* [23, 24]. The antibacterial and antifungal activities were also found in the BSs produced by *Halobacterium salinarum* [50]. It is quite fascinating the mechanism by which BS producers act. Various studies observed that the formation of a film on an interface after the excretion of a BS determines the attachment of certain microorganisms to the interface while inhibiting others. Therefore, it can be stated that some microorganisms can use their BS to regulate the cell surface properties in order to attach or to detach from surfaces according to need. In this respect, some authors [51] reported about studies on the mechanism by which bioemulsifier producing microorganisms regulate biofilm formation. Interestingly, the biofilm formation inhibition of *P. aeruginosa* ATCC10145 is highlighted and this seems to be determined to the BSs produced by a coral mucus associated strains [28]. According to [49], the detection of microorganisms able to produce BSs with such activities is fundamental for a reduced utilization of synthetic surfactants, and it favors the increase of biodegradable compounds. Besides, the existence of a horizontal transfer of high molecular weight emulsifiers from the producing microorganisms to heterologous bacteria was highlighted. In this case, the first step of this process is to bind to the surface of a group of bacteria by changing their surface properties in order to transport the emulsifier into the recipient cells. This has significant ecological implications for building a network of microbial BS producing strains in natural microbial communities. Last but not least, a new role for rhamnolipids in stimulation of plant and animal defense responses has emerged [5].

In particular, rhamnolipids have been demonstrated to have a direct biocide action on bacteria and fungi and to increase the susceptibility of certain Gram-positive and Gram-negative bacteria to specific antibiotics. Indeed, these biomolecules have been known as exotoxins produced by pathogens and are described as a new class of microbe-associated molecular patterns (MAMPs) that is, molecular signals which activate a large battery of defense-related genes of plants and animals by which a more specific immune response is determined [5, 52].

4. Methods and screening procedure for testing BS production

The screening procedure is constantly based on the performance of a selection of standard tests, differently chosen by authors with the attempt to carry out a fast and economic selective procedure. The BS chemical diversity is very wide and also different in their properties; thus, the screening procedure has to probe all the multifaceted activities, from the interfacial to the emulsifying, from the chelating to the foaming stabilization functions, and so on. The interfacial actions are generally explored by direct measurements of surface tension, through the evaluation of the force required to detach a ring or loop of wire (Du-Noüy method), or a platinum plate (Wilhelmy plate method) from an interface or surface [53, 54]. These methods ensure the advantages of accuracy and easiness, despite they require specialized equipment, and the impossibility to perform the measurements on different samples simultaneously. Other direct surface tension measurements have been reported, but they are actually considered not recommendable for an efficient screening procedure [55, 56]. The surface tension evaluation by direct measurement is one of the most commonly test reported for marine isolates [11, 15, 19, 20, 25, 26, 28, 47]. Many screening methods are based on indirect measure of surface/interfacial tension, such as drop collapse method, titled glass slide test, oil spreading assay, penetration assay, and microplate assay. The main advantages are represented by the possibility to screen more samples in a quick way, although a low sensitivity due to the strong dependence on BS concentration was highlighted. The methods are indeed based on distortion visual effects caused by the BS presence and are generally performed on supernatants and some authors suggested to color them in order to evidence the visual effects [57–59]. Among them, the oil spreading and the drop collapse assays are the most reported for marine bacteria [11–13, 21–24, 26, 28, 30, 50] for its rapidity and easiness, in addition to the requirement of small volume of sample, and simple and easily available equipment [56].

A second group of screening tests is differently based on the evaluation of emulsifying activity and is generally carried out with some modifications, with regards to volume of culture/supernatant, to the hydrocarbon used as test, and to vortexing time. The most used tests are the emulsifying activity test, based on a quick observation of emulsion occurrence [60] and the E_{24} index detection, based on the occurrence of emulsions stable over the time. It was applied in many screenings for marine BS producers [11–13, 19–25, 28, 30, 46–48, 50], and most authors reported some modifications: it was tested with cell broth instead of cell-free supernatant, whereas kerosene has been replaced in some cases by other hydrophobic compounds, for example, hexadecane, crude oil, vegetable oil, and diesel oil [13]. Surface activity and emulsification capacity do not always correlate [56]. Indeed, different studies [7, 14, 19] observed and explained this aspect considering that some BSs might stabilize (emulsifiers) or destabilize (de-emulsifiers) the emulsion so that the emulsification test alone fails to identify compounds with surfactant activity which destabilizes the emulsions. However, while the surface activity assay could give just an indication of BS production, the detection of stable emulsion index correlates to the surfactant concentration. According to the Refs. [20, 25], the surface tension measurement and the emulsification activity assays could be complementary and represent the basic tests to include in a screening procedure, allowing to detect both low molecular mass BSs with efficiency in surface and interfacial tension reduction and high molecular mass BSs more effective as emulsion stabilizers. In addition to the above standard screening tests, several authors reported the use of specific assays, the cetyltrimethylammonium bromide (C-TAB) agar plate assay, and the blood agar assay, useful to detect anionic BSs, but not enough sensitive to detect BS producers. The first one is based on the interaction between anionic surfactants eventually present and the cationic surfactant cetyltrimethylammonium bromide contained in a methylene blue stained mineral salts agar plate; the consequent creation of a blue dark halo around the colonies detects the presence of the BS producers. The blood agar assay is differently based on the hemolytic actions of biosurfactants—α, β, and γ hemolysis—on solid medium containing defibrinated blood as greenish or clarification halos around the bacterial colonies. The use of these specific assays have been reported in most of the above mentioned references, but it has been reduced over the years, because some authors signaled the possible harmfulness toward bacterial growth, the low specificity, and the possible occurrence of false positive/negative. In [19, 20], the two assays are performed on 69 and 96 isolates of different marine origin, respectively, suggesting the use of such tests as integration of a more deep screening procedure. To the list of screening test, it is necessary to mention some other tests with important implications in the screening quality and for considerations about bioremediation purposes. In [46], the authors reported the BS-mediated hydrocarbon degradation by *Joostella* sp. A8 grown in pure culture and consortia, as result of a BS production monitoring in which the bacterial adhesion to hydrocarbons assay (BATH assay) was included. This is a simple photometrical assay described for the first time in Ref. [61] for indirect evaluation of BS production by measuring the hydrophobicity of bacteria. The use of such test in relation to BS production is still a debating issue in this field, because the authors are divided between those who believes that a correlation between cell hydrophobicity, biodegradation efficiency, and BS production exists [62, 63] and who instead claims that the changes in cellular surface properties could be affected by several parameters and may not be necessarily associated to biodegradation ability [64]. Finally, some other minor tests have been reported by several authors for BS production by marine bacteria, but their efficiency have to be improved: this is the case of penetration assay [26], the hydrocarbon overlay agar method [13, 30], tilted glass slide test (TGS test) [30], and lipase activity [21–24, 28]. A number of additional screening tests are well reviewed in [59].

4.1 BS production conditions

There are two primary pathways for bacterial BS biosynthesis: the way of hydrocarbons and the way of carbohydrates [9, 14], and their production is influenced by the availability, types of carbon sources, and the balance between carbon-nitrogen and other limiting nutrients [10]. The effects of several parameters on marine bacterial BS production have been explored by several authors, with particular focus on carbon source, temperature, pH, and NaCl concentration [20, 22–27]. The bacterial BS production has been generally observed at early stationary phase of growth [21, 22, 44, 65], or at exponential phase [25, 27]. In Ref. [22], the authors hypnotized that *N. alba* releases a cell bound biosurfactant into the culture broth which leads to an increase in extra cellular BSs. Moreover, the same authors evidenced a good BS production by the strains at all the tested conditions, even if they detected for *Nocardiopsis alba* the optimum conditions for BS production at pH 7, temperature

30°C, and 1% salinity with glucose and peptone supplementation as carbon and nitrogen sources, respectively. In [21], the authors confirmed that carbon source and its concentration are affecting parameters for BS production, and established that glycerol, peptone, ferrous sulfate, and incubation time exhibited significant effect, with optimum levels as pH 7, temperature at 37°C, and salt concentration 2% for *B. amyloliquefaciens.* In particular, the authors reported that glycerol used as a carbon source showed the highest BS production (up to 6.76 g/l). According to Ref. [24], the studies performed a more deep optimization analysis to investigate the BS production by the sponge-associated *Brachybacterium paraconglomeratum* and indicated a yeast extract nitrogen source as factor enhancing up to 60% biosynthesis activity. Positive effects were also exhibited by the supplementation with 2% of NaCl, a pH level of 7.0, and a 30°C temperature. Moreover, asparagine resulted highly effective for BS production followed by glycine, leucine, and valine, and a requirement of $CuSO_4$ as a metal supplement was requested by the strain for optimum production of BS. In [20], the authors investigated the influence of salinity and temperature on the BS production by polychaete-associated isolates and showed that the NaCl concentration strongly influenced the surface tension reducing activity and emulsification rate in major level rather than temperature. Nevertheless, the authors reported that the strain *Marinobacter* sp. A1 exhibited the best performances at 15°C and in the absence of NaCl, by suggesting that limited conditions could act as stimulating factors. Interestingly, several researchers [19, 20, 25] also reported the BS production in the presence of hydrocarbons. The BS synthesis under solid state cultivation (SSC) was investigated for *Brevibacterium aureum* MSA13, which increased its production with pre-treated molasses, glucose, and acrylamide [23] as substrates. The report is interesting because it represents the first attempt in which acrylamide was used as nitrogen source, and the SSC conditions have been proven to be a preferred bioprocess for the BS production and optimization. In [27], the authors suggested a parallel relationship between bacterial growth and productivity and tested several carbon sources (sugar cane molasses, olive oil, corn oil, motor oil, and kerosene) among which molasses resulted the better one, as previously reported for non-marine BS producers [40]. This finding evidenced the possible importance of low cost substrate employment, which could solve the problems of high costs for BS production. As suggested by other researchers who revealed yeast extract and tryptone as significantly positive factors for BS production, nitrogen is an aspect to be carefully treated, as important constituent of the peptide part of lipopeptidic BSs [27]. Similarly, phosphate source has to be regulated to ensure a good bacterial growth, with positive influence on BS production. Finally, calcium source has also been reported as important positive factor and enhancer of emulsification activity. Due to this deep optimization approach, the authors achieved an increase of BS production with emulsification indexes from 60 to 77%.

5. Genetic regulation

The genetics of microbial surfactant synthesis is important because it represents a primary factor determining their productivity. It has been studied by the use of mutants naturally occurring or induced by transposition. However, the screening for such mutants is difficult because of the loss of ability to produce the surfactant that does not result in an easily selectable phenotype or may be lethal. The regulation at the molecular level of BS production has been mainly investigated for the rhamnolipid produced by *Pseudomonas aeruginosa* and for a few lipopeptides of Bacilli. The BS production has proved to be controlled by *quorum sensing* mechanism, a cell density dependent gene regulation process allowing bacterial cells to express certain specific genes on attaining high cell density [66]. With this mechanism, at the base of the bacterial production of secondary metabolites, bacteria produce a small diffusible signal molecule which accumulates in the growth medium and determine the gene activation when the microorganisms are in high densities. According to [49], the genes responsible for lipopeptide BS biosynthesis from *Bacillus* and *Pseudomonas* species display a high degree of structural similarity among themselves. The lipopeptide surfactin is produced as a result of nonribosomal biosynthesis. The mechanism is quite complex, as peptide synthetase for the amino acid moiety of surfactin is encoded by four open reading frames (ORFs) in the *Srf*A operon. The gene controlling the peptide synthetase consists of *Srf*AA (SrfORF1), *Srf*AB (SrfORF2), and *Srf*AD which activate each other. As a matter of fact, a more complex mechanism drives the *srf*A operon expression by controlling the level of various molecular signals [67]. On the other hand, structural genes required for the synthesis of the lipopeptidic BS lichenysin have been isolated, and a high sequence homology with *srf*A of surfactin synthetase was observed. Indeed, the lichenysin biosynthesis operon (called *lic operon*) was cloned and sequenced by revealing a composition of three peptide synthetase genes (lic A, lic B, and lic C) [68]. With respect to the genes which regulate the synthesis of glycolipids as rhamnolipids, they had been isolated, characterized, and their introduction into other species allowed the production of rhamnolipids in heterologous hosts [66]. The genes involved in the rhamnolipid biosynthesis are plasmid encoded and act during the late exponential early stationary phase as a consequence of the higher cell density. The synthesis of rhamnolipids in *P. aeruginosa* is carried out by the *rhl*AB operon, and a few additional genes are required [51, 45]. In particular, the biosynthetic pathway involved the genes *rhl*A, B, C, R, and I. *rhl*AB operon and *rhl*C gene encode the two rhamnosyltransferases (proteins that resides in the periplasm) responsible for the synthesis, transport or the stabilization of the rhamnosyltransferase within and in the cytoplasmatic membrane. *Rhl*A, B genes are organized in one operon and are coexpressed from the same promoter [67]. *Rhl*R and *rhl*I genes act as regulators of the *rhl*A, B gene expressions, and in turn are regulated by other genes (*las*R and *las*I) of a second quorum sensing system found in a different region of the *Pseudomonas aeruginosa* chromosome [67, 69]. As a matter of fact, the circuit of rhamnolipid production is promoted by other regulatory factors triggered by environmental conditions such as C/N ratio and inhibited by higher iron concentration [70], while the transcription of *rhl*AB genes is overexpressed under nitrogen-limited medium. As regards to the genetic regulation of high molecular weight heteropolysaccharide bioemulsifiers, they are more complex than the low molecular weight lipopeptides or rhamnolipids because they require a larger number of genes, and the genetic organization is even more complicated for polysaccharides-protein complexes [49]. However, although several structural and regulatory genes have been identified for the BS production, this aspect has been mainly improved for bacteria of terrestrial origin, while it has been scarcely explored for marine microorganisms. To the best of our knowledge, the only reports in this regard are in Refs. [23, 71]. In particular, these studies were addressed to the polyketide synthases (PKSs), the nonribosomal peptide synthases (NRPS), and large multifunctional proteins, modular proteins involved in the production of bioactive molecules. The research group investigated the possible correlation between the PKS gene and the BS biosynthesis in sponge-associated BS actinobacterial producers. In Ref. [23], the authors detected the presence of *pks* II gene in *B. aureum* MSA13, by suggesting the possible biosynthesis of secondary metabolites with antibiotic and surfactant properties. Furthermore, the studies described in [71] provided interesting insights into the KS genes of *Brevibacterium* and *Brachybacterium*, by confirming through molecular approaches that marine resources should be better explored for biodiscovery purposes. The molecular genetics of BS production is still in progress, and important genetic tools (plasmids, transposons, and gene libraries) are still to be developed, as well as further studies could allow to develop a biosynthetic engineering approach to design novel biomolecules.

6. Conclusions

The research on BS is still undergoing evolution and requires many improvements in several aspects. The results obtained in the last decades from the marine resources are very encouraging, both in terms of BS chemical diversity and in terms of the BS effectiveness and microbial production capacity. Despite this, new tools and improvements are necessary for a better comprehension of the genetic regulation, in order to exploit these mechanisms for a potential large-scale production. The discovery of new BSs must represent the main goal, and should be accompanied by appropriate chemical analysis, and identification of the most active fractions. The screening methods must be standardized and refined, and above all the study should be planned on the basis of the application of interest. Marine bioprospecting and the blue biotechnology are research areas that deserve to be explored, which are worth focusing on, and which could allow significant scientific contributions and useful applications for humans.

Author details

Rosanna Floris[1]*, Carmen Rizzo[2] and Angelina Lo Giudice[2,3]

1 AGRIS-Sardegna-Agricultural Research Agency of Sardinia, Servizio Ricerca Prodotti Ittici, Sassari, Italy

2 Department of Chemical, Biological, Pharmaceutical and Environmental Sciences, University of Messina, Messina, Italy

3 Institute for the Coastal Marine Environment, National Research Council, UOS Messina, Messina, Italy

*Address all correspondence to: rfloris@agrisricerca.it

References

[1] Baharum SN, Beng EK, Mokhtar MAA. Marine microorganisms: Potential application and challenges. Journal of Biological Sciences. 2010;**10**:555-564

[2] Dusane DH, Matkar P, Venugopalan VP, Kumar AR, Zinjarde SS. Cross-species induction of antimicrobial compounds, biosurfactants and quorum-sensing inhibitors in tropical marine epibiotic bacteria by pathogens and biofouling microorganisms. Current Microbiology. 2011;**62**:974-980

[3] Paniagua-Michel J, Olmos-Soto J, Morales-Guerrero ER. Algal and microbial exopolysaccharides: New insights as biosurfactants and bioemulsifiers. Advances in Food and Nutrition Research. 2014;**73**:221-257. (chapter eleven)

[4] Rodrigues L, Banat IM, Teixeira J, Oliveira R. Biosurfactants: Potential applications in medicine. Journal of Antimicrobial Chemotherapy. 2006;**57**:609-618

[5] Vatsa P, Sanchez L, Clement C, Baillieul F, Dorey S. Rhamnolipid biosurfactants as new players in animal and plant defense against microbes. International Journal of Molecular Sciences. 2010;**11**(12):5095-5108

[6] Rizzo C, Lo GA. Marine invertebrates: Underexplored sources of bacteria producing biologically active molecules. Diversity. 2018;**10**:52

[7] Chen C-Y, Baker SC, Darton RC. The application of high throughput analysis method for the screening of potential biosurfactants from natural sources. Journal of Microbiology Methods. 2007;**70**:503-510

[8] Batista SB, Mounteer AH, Amorim FR, Tòtola MR. Isolarion and characterization of biosurfactants/bioemulsifier-producing bacteria from petroleum contaminated sites. Bioresource Technology. 2006;**97**:868-875

[9] Desai JD, Banat IM. Microbial production of surfactants and their commercial potential. Microbiology and Molecular Biology Reviews. 1997;**61**:47-64

[10] Lotfabad TB, Shourian M, Roostaazad R, Najafabadi AR, Adelzadeh MR, Noghabi KA. An efficient biosurfactant-producing bacterium *Pseudomonas aeruginosa* MR01, isolated from oil excavation areas in south of Iran. Colloids and Surfaces B: Biointerfaces. 2009;**69**:183-193

[11] Dang NP, Landfald B, Willassen NP. Biological surface-active compounds from marine bacteria. Environmental Technology. 2016;**37**(9):1151-1158

[12] Dhail S, Jasuja ND. Isolation of biosurfactant-producing marine bacteria. African Journal of Environmental Science and Technology. 2012;**6**(6):263-266

[13] Shoeb E, Ahmed N, Akhter J, Badar U, Siddiqui K, Ansari FA, Waqar M, Imtiaz S, Akhtar N, Shaikh QA, Baig R, Butt S, Khan S, Khan S, Hussain S, Ahmed B, Ansari MA. Screening and characterization of biosurfactant-producing bacteria isolated from the Arabian Sea coast of Karachi. Turkish Journal of Biology. 2015;**39**:210-216

[14] Malavenda R, Rizzo C, Michaud L, Gerçe B, Bruni V, Syldatk C, Hausmann R, Lo Giudice A. Biosurfactant production by Arctic and Antarctic bacteria growing on hydrocarbons. Polar Biology. 2015;**38**:1565-1574

[15] Thavasi R, Sharma S, Jayalakshmi S. Evaluation of screening methods for

the isolation of biosurfactant producing marine bacteria. Journal of Petroleum & Environmental Biotechnology. 2011;**S-1**(1):1-6

[16] Pini F, Grossi C, Nereo S, Michaud L, Lo Giudice A, Bruni V, Baldi F, Fani R. Molecular and physiological characterisation of psychrotrophic hydrocarbon-degrading bacteria isolated from Terra Nova Bay (Antarctica). European Journal of Soil Biology. 2007;**43**:368-379

[17] Chrzanowski L, Ławniczak L, Czaczyk K. Why do microorganisms produce rhamnolipids? World Journal of Microbiology and Biotechnology. 2012;**28**:401-419

[18] Ławniczak L, Marecik R, Chrzanowski L. Contributions of biosurfactants to natural or induced bioremediation. Applied Microbiology and Biotechnology. 2013;**97**:2327-2339

[19] Rizzo C, Michaud L, Hörmann B, Gerçe B, Syldatk C, Hausmann R, De Domenico E, Lo Giudice A. Bacteria associated with Sabellids (Polychaeta: Annelida) as a novel source of surface active compounds. Marine Pollution Bulletin. 2013;**70**:125-133

[20] Rizzo C, Michaud L, Syldatk C, Hausmann R, De Domenico E, Lo Giudice A. Influence of salinity and temperature on the activity of biosurfactants by polychaete-associated isolates. Environmental Science and Pollutution Research. 2014;**21**:2988-3004

[21] Dhasayan A, Selvin J, Kiran S. Biosurfactant production from marine bacteria associated with sponge *Callyspongia diffusa*. Biotechnology. 2015;**5**:443-454

[22] Gandhimathi R, Kiran SG, Hema TA, Selvin J. Production and characterization of lipopeptide biosurfactant by a sponge associated marine actinomycetes *Nocardiopsis alba* MSA10. Bioprocess and Biosystems Engineering. 2009;**32**:825-835

[23] Kiran GS, Anto TT, Selvin J, Sabarathnam B. Optimization and characterization of a new lipopeptide biosurfactant produced by marine *Brevibacterium aureum* MSA13 in solid state culture. Bioresource Technology. 2010;**101**:2389-2396

[24] Kiran GS, Sabarathnam B, Thajuddin N, Selvin J. Production of glycolipid biosurfactant from sponge-associated marine actinobacterium *Brachybacterium paraconglomeratum* MSA21. Journal of Surfactants and Detergents. 2014;**17**:531-542

[25] Rizzo C, Syldatk C, Hausmann R, Gerçe B, Longo C, Papale M, Conte A, De Domenico E, Michaud L, Lo Giudice A. The demosponge *Halichondria* (Halichondria) *panicea* (Pallas, 1766) as a novel source of biosurfactant-producing bacteria. Journal of Basic Microbiology. 2018:1-11. https://doi.org/10.1002/jobm.201700669

[26] Graziano M, Rizzo C, Michaud L, Porporato EMD, De Domenico E, Spanò N, Lo Giudice A. Biosurfactant production by hydrocarbon-degrading *Brevibacterium* and *Vibrio* isolates from the sea pen *Pteroeides spinosum* (Ellis, 1764). Journal of Basic Microbiology. 2016;**56**:963-974

[27] Mabrouk MEM, Youssif EM, Sabry SA. Biosurfactant production by a newly isolated soft coral-associated marine *Bacillus* sp. E34: Statistical optimization and characterization. Life Science Journal. 2014;**11**:10

[28] Padmavathi AR, Pandian SK. Antibiofilm activity of biosurfactant producing coral associated bacteria isolated from Gulf of Mannar. Indian Journal of Microbiology. 2014;**54**:376-382

[29] Floris R, Scanu G, Fois N, Rizzo C, Malavenda R, Spanò N, Lo Giudice A. Intestinal bacterial flora of Mediterranean gilthead seabream (*Sparus aurata*, L.) as a novel source of natural surface active compounds. Aquaculture Research. 2018;**49**:1262-1273. DOI: 10111/are.13580

[30] Satpute SK, Banat IM, Dhakephalkar PK, Banpurkar AG, Chopade BA. Biosurfactants, bioemulsifiers and exopolysaccharides from marine microorganisms. Biotechnology Advances. 2010;**28**:436-450

[31] Mulligan CN. Environmental applications for biosurfactants. Environmental Pollution. 2005;**133**:183-198

[32] Rosenberg E, Ron EZ. High and low molecular mass microbial surfactants. Applied Microbiology and Biotechnology. 1999;**52**:154-162

[33] Pattanatu KSM, Gakpe E. Production, characterisation and applications of biosurfactants-Review. Biotechnology. 2008;**7**(2):360-370

[34] Rahman KSM, Street G, Lord R, Kane G, Rahman TJ, Marchant R, Banat IM. Bioremediation of petroleum sludege using bacterial consortium with biosurfactant. In: Singh SN, Tripathi RD, editors. Environmental Bioremediation Technologies. Springer Publication; 2006. pp. 391-408. DOI:10.1007/978-3-540-34793-4-17

[35] Reis CBLD, Morandini LMB, Bevilacqua CB, Bublitz F, Ugalde G, Mazutti MA, Jacques RJS. First report of the production of a potent biosurfactant with ß-trehalose by *Fusarium fujikuroi* under optimized conditions of submerged fermentation. Brazilian Journal of Microbiology. 2018;**S1517-8382**(17):30993-0

[36] Rosenberg E, Ron EZ. Surface active polymers from the genus *Acinetobacter*. In: Kaplan DL, editor. Biopolymers from Renewable Resources. New York: Springer, Berlin Heidelberg; 1998. pp. 281-291

[37] Desjardine K, Pereira A, Wright H, Matainaho T, Kelly M, Andersen RJ. Tauramamide, a lipopeptide antibiotic produced in culture by *Brevibacillus laterosporus* isolated from a marine habitat. Journal of Natural Products. 2007;**70**:1850-1853

[38] Kalinovvskaya NI, Kuznetsova TA, Rashkes YV, Milgrom YM, Milgrom EG, Willis RH. Surfactin-like structures of five cyclic depsipeptides from the marine isolate of *Bacillus pumillus*. Russian Chemical Bulletin. 1995;**44**(5):951-955

[39] Gerard J, Lloyd R, Barsby T, Haden P, Kelly MT, Andersen RJ. Massetolides A-H, antimycobacterial cyclic depsipeptides produced by two pseudomonads isolated from marine habitats. Journal of Natural Products. 1997;**60**(3):223-229

[40] Joshi SJ, Suthar H, Yadav AK, Hinguaro K, Nerurkar A. Occurrence of biosurfactants producing *Bacillus* spp. in diverse habitats. 2013 International Scholarly Research Network ISRN Biotechnology. Article ID 652340. p. 6. DOI: 10.5402/2013/652340

[41] Yakimov MM, Timmis KN, Wray V, Fredrickson HL. Characterization of a new lipopeptide surfactant produced by thermotolerant and halotolerant subsurface *Bacillus licheniformis* Bas50. Applied and Environmental Microbiology. 1995;**61**:1706-1713

[42] Gutierrez T, Shimmied T, Haidon C, Black K, Green DH. Emulsifying and metal ion binding activity of a glycoprotein exopolymer produced by *Pseudoalteromonas* sp. Strain TG12. Applied and Environmental Microbiology. 2008, 2008;**74**:4867-4876

[43] Kumar AK, Mody K, Jha B. Bacterial exopolysaccharides—A perception. Journal of Basic Microbiology. 2007;**47**:103-117

[44] Zhang Y, Miller RM. Enhanced octadecane dispersion and biodegradation by a *Pseudomonas* rhamnolipid surfactant (biosurfactant). Applied and Environmental Microbiology. 1992;**58**:3276-3282

[45] Perfumo A, Smyth TJP, Marchant R, Banat IM. Production and roles of biosurfactants and bioemulsifiers in accessing hydrophobic substrates. In: Timmis KN, editor. Handbook of Hydrocarbon and Lipid Microbiology. Chichester: Springer-Verlag Berlin Heidelberg; 2010. pp. 1502-1510. DOI: 10.1007/978-3-540-77587-4_103

[46] Rizzo C, Rappazzo AC, Michaud L, De Domenico E, Rochera C, Camacho A, Lo Giudice A. Efficiency in hydrocarbon degradation and biosurfactant production by *Joostella* sp. A8 when grown in pure culture and consortia. Journal of Environmental Sciences. 2018;**67**:115-126

[47] Rizzo C, Michaud L, Graziano M, De Domenico E, Syldatk C, Hausmann R, Lo Giudice A. Biosurfactant activity, heavy metal tolerance and characterization of *Joostella* strain A8 from the Mediterranean polychaete *Megalomma claparedei* (Gravier, 1906). Ecotoxicology. 2015;**24**:1294-1304

[48] Rizzo C, Lo Giudice A. Heavy metal tolerance and chelating activity of bacteria associated with Mediterranean polychaetes. SF Journal of Environmental and Earth Science. 2018;**1**(2):1015

[49] Das P, Mukherjee S, Sen R. Genetic regulations of the biosynthesis of microbial surfactants: An overview. Biotechnology & Genetic Engineering Reviews. 2008;**25**(1):165-186

[50] Sumaiya M, Anchana Devi C, Leela K. A study on biosurfactant production from marine bacteria. International Journal of Scientific and Research Publications. 2017;**7**:12

[51] Ron EZ, Rosenberg E. Natural roles of biosurfactants. Environmental Microbiology. 2001;**3**(4):229-236

[52] Mackey D, McFall AJ. MAMPs and MIMPs: Proposed classifications for inducers of innate immunity. Molecular Microbiology. 2006;**61**:1365-1371

[53] Tadros T. Adsorption of surfactants at the air/liquid and liquid/liquid interfaces. In: Tadros TF, editor. Applied Surfactants: Principles and Applications. Weinheim: Wiley VCH; 2015. pp. 81-82

[54] Tuleva B, Christova N, Jordanov B, Jordanov B, Nikolova-Damyanova B, Petrov P. Naphthalene degradation and biosurfactant activity by *Bacillus cereus* 28BN. Zeitschrift für Naturforschung. Section C. 2005;**60**:577-582

[55] Dilmohamud B, Seeneevassen J, Rughooputh S, Ramasami P. Surface tension and related thermodynamic parameters of alcohols using the Traube stalagmometer. European Journal of Physics. 2005;**26**(6):1079-1084

[56] Plaza G, Zjawiony I, Banat I. Use of different methods for detection of thermophilic biosurfactant-producing bacteria from hydrocarbon-contaminated bioremediated soils. Journal of Petroleum Science and Engineering. 2006;**50**(1):71-77

[57] Chen C, Baker S, Darton R. The application of a high throughput analysis method for the screening of potential biosurfactants from natural sources. Journal of Microbiological Methods. 2007;**70**:503-510

[58] Tugrul T, Cansunar E. Detecting surfactant-producing microorganisms

by the drop-collapse test. World Journal of Microbiology and Biotechnology. 2005;**21**:851-853

[59] Walter V, Syldatk C, Hausmann R. Screening concepts for the isolation of biosurfactant producing microorganisms. Advances in Experimental Medicine and Biology. 2010;**672**:1-13

[60] Christova N, Tuleva B, Lalchev Z, Jordanovac A, Jordanov B. Rhamnolipid biosurfactants produced by *Renibacterium salmoninarum* 27BN during growth on n-hexadecane. Zeitschrift für Naturforschung. Section C. 2004;**59**(1-2):70-74

[61] Rosenberg M, Gutnick D, Rosenberg E. Adherence of bacteria to hydrocarbons—A simple method for measuring cell-surface hydrophobicity. FEMS Microbiology Letters. 1980;**9**(1):29-33

[62] Franzetti A, Caredda P, Colla PL, Pintus M, Tamburini E, Papacchini M, Bestetti G. Cultural factors affecting biosurfactant production by *Gordonia* sp. BS29. International Biodeterioration and Biodegradation. 2009;**63**:943-947

[63] Obuekwe CO, Al-Jadi ZK, Al-Saleh ES. Insight into heterogeneity in cell-surface hydrophobocity and ability to degrade hydrocarbons among cells of two hydrocarbon-degrading bacterial populations. Canadian Journal of Microbiology. 2007;**53**:252-260

[64] Chakraborty S, Mukherji S, Mukherji S. Surface hydrophobicity of petroleum hydrocarbon degrading *Burkholderia* strains and their interactions with NAPLs and surfaces. Colloids and Surfaces. B, Biointerfaces. 2010;**78**:101-108

[65] Antoniou E, Fodelianakis S, Korkakaki E, Kalogerakis N. Biosurfactant production from marine hydrocarbon-degrading consortia and pure bacterial strains using crude oil as carbon source. Frontiers in Microbiology. 2015;**6**:274

[66] Daniels R, Vanderleyden J, Michiels J. Quorum sensing and swarming migration in bacteria. FEMS Microbiology. 2004;**28**:261-289

[67] Sullivan ER. Molecular genetics of biosurfactants production. Current Opinion in Biotechnology. 1998;**9**(3):263-269

[68] Yakimov MM, Kroeger A, Slepak TN, Giuliano L, Timmis KN, Golyshin PN. A putative lichenysis A Synthetase operon in *Bacillus licheniformis*: Initial characterization. Biochimica et Biophysica Acta. 1998;**1399**:141-153

[69] Ochsner UA, Fiechter A, Reiser J, Witholt B. Production of *Pseudomonas aeruginosa* rhamnolipid biosurfactant in heterologous hosts. Applied and Environmental Microbiology. 1995;**61**:3503-3506

[70] Guerra-Santos LH, Kappel O, Fiechter A. Dependence of *Pseudomonas aeruginosa* biosurfactant production in continuous culture biosurfactant production on nutritional and environmental factors. Applied Microbiology and Biotechnology. 1986;**24**:443-448

[71] Selvin J, Sathiyanarayanan G, Lipton AN, Al-Dhabi NA, Arasu MV, Kiran GS. Ketide synthase (KS) domain prediction and analysis of iterative type II PKS gene in marine sponge-associated actinobacteria producing biosurfactants and antimicrobial agents. Frontiers in Microbiology. 2016;**7**:63

Chapter 2

Fecal Metabolomics Insights of Agavins Intake in Overweight Mice

Alicia Huazano-García, Horacio Claudio Morales-Torres, Juan Vázquez-Martínez and Mercedes G. López

Abstract

Targeted and non-targeted metabolite profiling can identify biomarkers after a dietary treatment leading to a better understanding of interactions between diet and health. This study was conducted to establish enriched or depleted metabolites in the feces of overweight mice after a diet shift plus agavins or inulins supplementation, and their possible association with beneficial effects on host health. Thirty-eight male C57BL/6 mice were fed with a high-fat diet for 5 weeks followed by a diet shift to a standard diet supplemented with agavins (HF-ST + A) or inulins (HF-ST + I) for five more weeks. Feces were collected before and after prebiotic supplementation for metabolomics analyses. HF-ST + I group increased the fecal excretion of two methyl esters: linoleic and oleic acid, while HF-ST + A mice showed a substantial augment of 2-decenal, fructose, cyclohexanol, and the acids: 10-undecenoic, 3-phenyllactic, nicotinic, 5-hydroxyvaleric, and lactic. From the metabolites identified in HF-ST + A, only lactic acid has been reported previously and associated with beneficial effects on host health. However, the identification of new metabolites, coming from the microbial fermentation of agavins, opens opportunities to transform this information into practical solutions to tackle overweight and associated metabolic syndrome.

Keywords: agavins, branched neo-fructans, metabolomics, postbiotics, prebiotics, overweight, fecal metabolites, biomarkers

1. Introduction

In the last decade, an increasing number of studies have been strongly associated with a high-fat consumption altering the gut microbiota composition and/or its functionality [1, 2]. It has also been related to overweight and obesity as well as the metabolic syndrome [3–5]. Overweight and obesity not only affects the wellbeing of an individual but also places an unwanted economic burden on society [6]. Therefore, it is necessary and urgent to find an effective way to prevent and/or treat these worldwide pathologies. In this sense, prebiotics might be a good nutritional alternative in the management of overweight and obesity and its associated metabolic syndrome, since their supplementation or consumption can modulate the gut microbiota producing a wide range of metabolites (postbiotics), consequently generating positive effects on host health [6, 7].

Agavins are relatively new prebiotics that in pre-clinical studies have shown several beneficial effects on the health of individuals [8–10]. Agavins are neo-fructans composed of complex and highly branched molecules with β(2–1) and β(2–6) linkages as well as an internal glucose unit [11, 12].

Our research group has evidenced that agavins can decrease glucose, triglycerides, and cholesterol concentrations as well as increase the anorexigenic peptide glucagon-like peptide1 (GLP-1; appetite-suppressing peptide) secretion on mice fed with a normal diet [8, 9]. Moreover, recently, we reported that agavins intake led to the reversion of metabolic disorders in overweight mice (induced by high-fat diet consumption) and also a substantial decrement of orexigenic peptide ghrelin (appetite-stimulating) and adipokines (leptin and insulin) levels in the portal vein, in such a way that all mice showed an integral improvement on their health [10].

On the other hand, due to the structural complexity of agavins, they cannot be degraded by endogenous gastrointestinal enzymes during their passage through the stomach and the small intestine; so, they reach the colon structurally unchanged, where they are fermented by the gut microbiota present in this organ [13, 14]. Fermentation of complex carbohydrates, such as agavins, might involve the collaboration of a highly diverse selection of gut microbes, which produce a myriad of different metabolites (postbiotics) that are suggested as key links in the communication between bacterial communities of the gut and the host [15, 16]. Nonetheless, only short-chain fatty acids (SCFA) such as acetate, propionate, butyrate, lactate, and succinate are among the metabolites reported up to now, derived from the agavins fermentation in *in vitro* and/or *in vivo* studies [9, 10, 17, 18]; the generation of these acids has been associated with different beneficial effects in the context of obesity, since SCFA reduce body weight gain, through G-protein-coupled receptors (GPRs), influencing the secretion of hormones involved in appetite control [19–21]. Moreover, SCFA are used as energy sources and may contribute to several metabolic pathways, including gluconeogenesis [22] and lipogenesis [23], thus contributing to whole-body energy homeostasis.

In spite of the above, other secondary metabolic products from the microbiota such as amino acids, nucleotides, bile acids, phenolic acids, fatty acids, and sterols, to mention some, can come from the agavins-microbiota interactions that have yet to be established. In the last decades, developed metabolomic tools have allowed researchers to study and characterize a wide range of metabolites in a non-invasive manner and also on biological systems, obtaining a large set of metabolites (metabolomics) that derive from gut microbes, enriched or depleted, after a dietary intervention [24]. This area of studies has been increased on the last decade, since this opens an opportunity to propose new biomarkers with new therapeutic approaches, through selective alterations of microbial production molecules to promote host health and prevent diseases [25].

In the present work, we established general and unique metabolites in overweight mice after agavins intake. We have previously showed that agavins consumption by overweight mice led to a gut microbiota modulation (these changes differed from those originated by inulins intake [14]); then, we hypothesized that agavins structure and the changes in the composition of gut microbiota in relation to inulins could lead to changes in colonic metabolic activity. The identification of microbial metabolites derived exclusively from agavins consumption may help to propose new biomarkers with huge potential and applicability on the prevention and/or treatment of overweight and their comorbidities.

2. Materials and methods

2.1 Animals and diets

Thirty-eight male C57BL/6 mice (12 weeks old at the beginning of the experiment were obtained from Universidad Autonoma Metropolitana, Mexico city, Mexico) and housed in a temperature and humidity controlled room with a 12-h light–dark cycles. Mice were maintained in individual cages and subject to two experimental phases, to gain and loss weight, respectively. In the first phase, mice were fed with a high-fat diet (n = 30; 58Y1 Test Diet, St. Louis, MO, USA) for 5 weeks to induce overweight in the animals. In the second phase, overweight mice (HF) were shifted to the standard diet (5053 Lab Diet, St. Louis, MO, USA) alone (HF-ST; n = 8) or supplemented with agavins (HF-ST + A; n = 8) or inulins (HF-ST + I; n = 8) for five more weeks. Moreover, we had a healthy control group of mice (ST; n = 8), which were fed with the standard diet (5053 Lab Diet, St. Louis, MO, USA) throughout the experiment.

The high-fat diet (58Y1 Test Diet) had 20.3% calories from carbohydrates (16.15% maltodextrin, 8.85% sucrose, and 6.46% powdered cellulose), 18.1% from proteins, and 61.6% from fat (31.7% lard and 3.2% soybean oil), whereas the standard diet (5053 Lab Diet) contained 62.4% calories from carbohydrates (28.6% starch, 3.24% sucrose, 1.34% lactose, 0.24% fructose, and 0.19% glucose), 24.5% from proteins, and 13.1% from fat.

Food and water were provided *at libitum* along the experiment. Mice experiments were conducted according to the Mexican Norm NOM-062-ZOO-1999 and approved by the Institutional Care and Use of Laboratory Animals Committee from Cinvestav-Mexico (CICUAL; protocol number 0091-14).

2.2 Agavins and inulins

Agavins from 4-year-old *Agave tequilana* Weber blue variety plants were extracted and purified in our laboratory and presented an average degree polymerization (DP) of 8, whereas inulins (oligofructose) was bought from Megafarma® (Mexico city, Mexico) and possess an average DP of 5. Agavins and inulins were added in the water at a concentration of 0.38 g/mouse/day [10].

2.3 Feces collection and preparation for untargeted and organic acids metabolic analyses

Feces were collected from each mouse at the end of the first and second experimental phases, before and after prebiotic supplementation, respectively. The feces of mice were pooled by treatment, lyophilized, triturated, and homogenized to generate fecal metabolites profiles. Untargeted metabolic analysis was carried out following a method adjusted from Eneroth et al. [26] and Gao et al. [27] as follows: 100 mg of feces were extracted tree times with chloroform/methanol (2:1), 1 mL each time. After that, the extracts were combined and solvent freed. The residue was resuspended in 1 mL of chloroform/methanol (2:1) and an aliquot of 50 μL was transferred to a vial. The aliquot was solvent freed under nitrogen flux and then was derivatized using BSTFA with 1% TCMS (80 μL) and pyridine (20 μL) at 80°C for 25 min. Once the system was at room temperature, isooctane was added to a final volume of 200 μL. Heptadecanoic acid, at final concentration of 3 mg/mL, was used as internal standard.

On the other hand, extraction of organic acids was performed according to García-Villalba et al. [28]. Briefly, 100 mg of feces were suspended in 1 mL of aqueous 0.5% phosphoric acid solution and mixed in vortex for 2 min. After that, samples were centrifuged for 10 min at 10,000 g. Then, the supernatant was transferred to a vial and was extracted with an equal volume of ethyl acetate. 2-Methyl valeric acid was used as internal standard at final concentration of 2 mM. This system was centrifuged for 10 min at 10,000 g, and then 200 μL of the ethyl acetate phase were transferred to a vial, dried under nitrogen flux, and derivatized using 80 μL of BSFTA with 1% TMCS and 20 μL of pyridine. The mix was allowed to react at 80°C for 25 min. After the mix was at room temperature, isooctane was added to a final volume of 200 μL.

2.4 Gas chromatography/mass spectrometry analysis

For GC/MS analysis, 1 μL of the organic phase was injected in the pulsed-splitless mode. Injector temperature was set to 260°C. A HP-5-MS capillary column (30 m × 25 μm × 0.25 μm) was used with helium as carrier gas at constant flow rate of 1 mL/min. Oven program began at 40°C (held 5 min), then increased at rate of 6°C/min until 170°C, then a second temperature ramp of 12°C/min until 290°C was applied. Transfer line temperature was set at 260°C. Mass spectrometer operated at 70 eV of electron energy, quadrupole and ion-source temperatures were 150 and 230°C, respectively. Data were obtained scanning from 40 to 550 m/z, while MassHunter Workstation software version B.0.0.6 (Agilent Technologies, Inc.) was used to collect the data. Components mass spectra and retention times were obtained using the AMDIS (automated mass spectral deconvolution and identification system, http://www.amdis.net/) software. Compounds identification was achieved by comparing their respective extracted mass spectrum with the mass spectra of the standards and/or with the mass spectra data of the NIST library and software (National Institute of Standards and Technology, USA).

2.5 Statistics and data analysis

Results are present as mean ± standard deviation. Differences between the diets were determined using one-way ANOVA followed by a Tukey post hoc test or a Dunett T3 post hoc test. Differences were considered significant when $p < 0.05$. Statistical analyses were performed using the IBM SPSS Statistics software version 22. Principal component analysis and heatmap were conducted using a language and environment for statistical computing R and the ade4 and gplots packages.

3. Results and discussion

Previous studies carried out by our research group evidenced that agavins intake led to improvement on health and wellness of the host, which has been associated with gut microbiota modulation and their metabolic products such as SCFA [10, 14]; nonetheless, other bioactive chemical compounds coming from agavins fermentation that could also contribute with the beneficial agavins consumption effects are unknown yet. In the present work, we performed a metabolomics analysis to establish and propose general and unique metabolites (postbiotics) in the feces of overweight mice after agavins (prebiotic) intake. Since it has been recommended the use of combination of methodologies to extend the metabolic coverage of microbiota [29], we performed an untargeted metabolic as well as organic acids profile analyses to carry out this task.

3.1 Untargeted metabolic profiles

Untargeted metabolic analysis showed a total of 300 metabolites, from which only 109 were identified completely. Those 109 compounds mainly included fatty acids and their esters, carbohydrates, sterols, alcohols, alkanes, SCFA, aldehydes, amino acids, nucleobases, bile salts, and phenylpropanoids (**Table 1**).

Of these 109 compounds (**Table 1**), 32 presented a significantly differential abundance between the different evaluated diets (Tukey's test, $p < 0.05$; **Table 2**). These 32 metabolites were grouped mostly in fatty acids and sterols and subsequently used for the principal component analysis (PCA) and heat map construction.

PCA was applied to the data to investigate metabolomics changes derived from agavins consumption. The variance explained by each principal component (PC) is displayed on the X and Y axes. PC1 and PC2 account for 62.5 and 20.5% of the variance, respectively. Very clear and separated clusters were observed among overweight mice (HF) and the other mice groups, suggesting differences on fecal metabolomics profiling (**Figure 1A**). In addition, mice that received agavins supplement are clearly separate, into a distinct cluster, from rodents fed with inulins supplementation. Interestingly, HF-ST + A group appear very close to standard diets (HF-ST and ST), evidencing a large similarity on metabolites among them compared to HF-ST + I group (**Figure 1A**).

Moreover, the loading plot illustrates the variables (metabolites) that are responsible for the discrimination (clustering of the samples) observed in the PCA plot. Then, according to the loading plot, HF and HF-ST + I groups shared the PC2 due to high content of oleic acid, cholesterol, and stigmasterol in the feces (**Figure 1B**).

ID	RT	Compound	Family
1	12.62	Cyclohexanol	Alcohol
2	13.028	Carbonic acid	Others
3	13.183	β-Hydroxybutyric acid	βOH-SCFA
4	13.234	Heptanoic acid	FA
5	13.281	α-Hydroxyvaleric acid	αOH-SCFA
6	14.008	Benzaldehyde, 2,5-dimethyl-	Aldehyde
7	14.049	L-Valine	Amino acid
8	14.483	Urea	Others
9	14.623	2-Decenal, (E)-	Aldehyde
10	14.94	Glycerol	Polyalcohol
11	15.074	2,5-dihydroxy hexane, 2,5-dimethyl-	Polyalcohol
12	15.406	Succinic acid	DiCAc
13	15.778	Uracil	Nucleobase
14	16.153	Butane, 1,2,4, triol	Polyalcohol
15	16.646	Thymine	Nucleobase
16	16.774	Hydrocinnamic acid	PhePr
17	17.096	Bicyclo[2.2.1]heptane-1-carboxylic acid, 7-Hydroxy, methyl ester	Ester
18	17.258	Decanoic acid	FA
19	17.55	Decane, 2,3,5,8-tetramethyl-	Alkane

ID	RT	Compound	Family
20	17.666	Dodecane, 4,6-dimethyl-	Alkane
21	17.741	Dodecane, 2,6,11-trimethyl-	Alkane
22	17.95	2,4-Ditert-butylphenol	Phenolic
23	18.011	1-Butene, 1-phenyl-3-(hydroxy)-, E	Ar
24	18.168	Pyroglutamic acid	Amino acid
25	18.422	2,6-Ditert-butylphenol	Phenolic
26	18.642	Pentadecane	Alkane
27	18.637	Undecanoic acid	FA
28	18.724	Heptadecane, 2,6,10,15-tetramethyl-	Alkane
29	19.173	*m*-Hydroxyphenylacetic acid	OH-Ar-SCFA
30	19.228	Cyanuric acid	Triazine
31	19.428	D-Arabinose	CHO
32	19.52	*p*-Hydroxyphenylacetic acid	OH-Ar-SCFA
33	19.599	*n*-Dodecanoic acid	FA
34	20.172	Hexadecane, 2,6,11,15-tetramethyl-	Alkane
35	20.204	Xylulose	CHO
36	19.153, 20.341	D-Mannose	CHO
37	20.438	10-Undecenoic acid	FA
38	20.51	Benzenepropanoic acid	PhePr
39	21.051	Glycerol phosphate	Others
40	21.159	Tetradecanoic acid, 12-methyl-, methyl ester	FAME
41	21.225	Azelaic acid	DiCAc
42	19.549, 21.389	D-Galactose	CHO
43	21.583, 21.652	D-Fructose	CHO
44	21.345	Tetradecanoic acid	FA
45	21.878	Inositol	Polyalcohol
46	22.039	Adenine	Nucleobase
47	22.352, 22.436	Methyl-tetradecanoic acid isomers (C15 fatty acid isomers)	FA
48	22.716	Pentadecanoic acid	FA
49	21.966, 22.483, 23.336	D-Glucose	CHO
50	23.428	*cis*-9-Hexadecenoic acid	FA
51	23.481	*trans*-9-Hexadecenoic acid	FA
52	23.691	Hexadecanoic acid	FA
53	24.128	Linoleic acid, methyl ester	FAME
54	24.177	Oleic acid, methyl ester	FAME
55	24.37	*cis*-10-Heptadecenoic acid	FA
56	24.399	Stearic acid, methyl ester	FAME
57	24.404	α-D-glucose, 2-(acetylamino)-2-deoxy	CHO
58	25.102	5-Hydroxyindoleacetic acid	IndolAc

ID	RT	Compound	Family
59	25.198	Linoleic acid	FA
60	25.25	Oleic acid	FA
61	25.298	*trans*-11-Octadecenoic acid	FA
62	25.315	*cis*-11-Octadecenoic acid	FA
63	25.481	Stearic acid	FA
64	26.262	2-O-glycerol-α-D-galactose	CHO
65	26.288	Nonadecanoic acid	FA
66	26.546	Arachidonic acid	FA
67	26.648	tert-Hexadecanethiol	Thiol
68	26.716	Tetratriacontane	Alkane
69	26.743	Octadecane, 3-ethyl-5-(2-ethylbutyl)-	Alkane
70	26.774	Succinylacetone	Others
71	26.838	Hentriacontane	Alkane
72	26.9	Oleamide	Amide
73	27.006	Sebacic acid	DiCAc
74	27.095	Eicosanoic acid	FA
75	27.509	1-O-hexadecylglycerol	Glycerol-ether
76	27.584	Heneicosanoic acid	FA
77	27.618	Propyl myristate	Ester
78	27.716	2-Octadecenoic acid	FA
79	27.866	Heneicosanoic acid	FA
80	28.498	4-n-octadecylcyclohexane, 1,3,5-trimethyl-	Alkane
81	28.62	Docosanoic acid	FA
82	28.76	α-Hydroxy sebacic acid	αOH-DiCAc
83	28.967	1-O-Octadecylglycerol	Glycerol-ether
84	29.01	*cis*-4-Tetradecene, 2-methyl-	Alkene
85	29.339	Tricosanoic acid	FA
86	29.578	1-Monooleoylglycerol	Monoglyceride
87	29.709	1-Docosanol	Alcohol
88	29.895	*cis*-15-Tetracosenoic acid	FA
89	28.125, 28.661, 29.646, 29.932	Disaccharides (including sucrose)	CHO
90	30.026	Octadecane, 3-ethyl-5-(2-ethylbutyl)-	Alkane
91	30.221	Enterolactone	Lignin
92	30.329	1-O-hexadecylglycerol	Glycerol-ether
93	30.383	Tricosanol	Alcohol
94	30.428	Cholesta-2,4-diene	Sterol
95	30.673	Cholesta-3,5-diene	Sterol
96	31.31	β-Tocopherol	Vitamin
97	31.454	Hexacosanoic acid	FA
98	31.891	Coprostan-3-ol	Sterol

ID	RT	Compound	Family
99	32.551	α-Tocopherol	Vitamin
100	32.768	Cholesterol	Sterol
101	33.336	Lanosterol	Sterol
102	33.852	Campesterol	Sterol
103	34.207	Stigmasterol	Sterol
104	34.346	Chenodeoxycholic acid	Bile salt
105	34.563	Xi-Ergost-7-ene, 3β-	Sterol
106	34.918	β-Sitosterol	Sterol
107	35.064	24-Ethylcoprostanol	Sterol
108	35.268	*trans*-Dehydroandrosterone	Steroid
109	38.518	Urs-12-en-28-al, 3-(acetyloxy)-, (3β)-	Sterol

Some metabolites have more than one retention time (RT) due to the presence of isomers. SCFA, short-chain fatty acid; FA, fatty acid; FAME, fatty acid methyl ester; CHO, carbohydrate; αOH, alfa-hydroxy; βOH, beta-hydroxy; DiCAc, dicarboxylic acid; PyrCAc, pyridine carboxylic acid; ωOH, omega-hydroxy; Cy, cyclic; Ar, aromatic; PhePr, phenylpropanoid; TCA, tricarboxylic acid; IndolAc, indolic acid.

Table 1.
Metabolites identified in the feces of mice.

ID	RT	Compound	Family	Fold change			
				HF	HF-ST	HF-ST + A	HF-ST + I
1	12.62	Cyclohexanol	Alcohol	−1.00	0.10	0.57	−0.13
3	13.183	β-Hydroxybutyric acid	βOH-SCFA	−0.62	−0.48	0.15	0.34
4	13.234	Heptanoic acid	FA	−0.88	−0.39	−0.13	−0.13
5	13.281	α-Hydroxyvaleric acid	αOH-SCFA	−0.30	−0.18	0.56	1.10
9	14.623	2-Decenal, (E)-	Aldehyde	−1.00	0.18	1.79	−0.90
10	14.94	Glycerol	Polyalcohol	−0.90	−0.15	−0.16	0.01
12	15.406	Succinic acid	DiCAc	−1.00	−0.27	1.07	2.25
16	16.774	Hydrocinnamic acid	PhePr	−1.00	−0.15	−0.30	0.05
22	17.95	2,4-Ditert-butylphenol	Phenolic	−0.40	0.39	0.52	−0.16
24	18.168	Pyroglutamic acid	Amino acid	−0.78	0.00	0.17	1.29
32	19.52	*p*-Hydroxyphenylacetic acid	Ar-acid	−0.75	−0.27	−0.36	0.19
37	20.438	10-Undecenoic acid	FA	−0.75	0.01	1.03	−0.68
38	20.51	Benzenepropanoic acid	PhePr	−0.95	−0.28	−0.32	−0.24
39	21.051	Glycerol phosphate	Others	−0.98	−0.07	−0.16	0.24
40	21.159	Tetradecanoic acid, 12-methyl, methyl ester	FAME	−0.31	0.02	0.58	2.27
43	21.583 21.652	D-Fructose	CHO	−0.93	−0.88	0.36	−0.57
44	21.345	Tetradecanoic acid	FA	−0.57	0.12	0.01	0.18
46	22.039	Adenine	Nucleobase	−0.92	−0.34	0.08	0.36
48	22.716	Pentadecanoic acid	FA	−0.68	−0.12	−0.04	0.48

ID	RT	Compound	Family	Fold change			
				HF	HF-ST	HF-ST + A	HF-ST + I
49	21.966 22.483 23.336	D-Glucose	CHO	−0.98	−0.40	0.02	0.47
53	24.128	Linoleic acid, methyl ester	FAME	−1.00	0.02	0.01	2.94
54	24.177	Oleic acid, methyl ester	FAME	−1.00	−0.07	−0.15	1.90
56	24.399	Stearic acid, methyl ester	FAME	−0.60	−0.15	0.02	1.15
58	25.102	5-Hydroxyindoleacetic acid	IndolAc	−1.00	0.57	0.86	1.38
59	25.198	Linoleic acid	FA	−0.71	−0.12	−0.41	0.34
60	25.25	Oleic acid	FA	0.34	−0.09	−0.35	0.18
81	28.62	Docosanoic acid	FA	−0.55	−0.01	−0.03	0.25
98	31.891	Coprostan-3-ol	Sterol	−0.08	0.21	0.54	1.53
100	32.768	Cholesterol	Sterol	0.52	−0.09	−0.35	0.27
103	34.207	Stigmasterol	Sterol	−0.13	−0.17	−0.47	0.04
106	34.918	β-Sitosterol	Sterol	−0.75	−0.14	−0.44	−0.04
107	35.064	24-Ethylcoprostanol	Sterol	−0.86	−0.12	−0.29	−0.09

Fold change value was calculated by comparison with the healthy mice fed with a standard diet (ST). HF, overweight mice; HF-ST, overweight mice that were switched to a standard diet; HF-ST + A, overweight mice changed to standard diet plus agavins; HF-ST + I, overweight mice changed to standard diet plus inulins. All the metabolites listed here have significant difference at least in one treatment $p < 0.5$. ID numbers correspond with those of ***Table 1****. SCFA, short-chain fatty acid; FA, fatty acid; FAME, fatty acid methyl ester; CHO, carbohydrate; αOH, alfa-hydroxy; βOH, beta-hydroxy; DiCAc, dicarboxylic acid; Ar, aromatic; PhePr, phenylpropanoid; IndolAc, indolic acid.*

Table 2.
Fold-change of differential metabolites detected in the feces of overweight mice after a diet switch and prebiotic supplementation.

Besides, hierarchical clustering analysis (**Figure 2**) revealed that HF group had a very low content of most identified metabolites, with exception of cholesterol and oleic acid. In addition, the bile salt chenodeoxycholic acid was detected exclusively in the HF treatment, and although it was not included in the hierarchical analysis since it was not detected in any other treatment, this metabolite could be used a biomarker for HF diet.

Interestingly, only HF-ST + I mice presented a high content of cholesterol and oleic acid in its feces; therefore, in the heatmap, this group appears very close to HF (**Figure 2**). Moreover, HF-ST treatment is closely linked to ST group due to similar content of all evaluated compounds. While, HF-ST + A group is located as the link between HF-ST + I and the cluster of standard diets (HF-ST and ST **Figure 2**).

In contrast to HF-ST group, HF-ST + A and HF-ST + I exhibited an enrichment of the following acids: succinic, β-hydroxybutyric (BHB), α-hydroxyvaleric, and pyroglutamic; as well as 12-methyl-tetradecanoic acid methyl ester and adenine, which could be used as biomarkers for mice groups with prebiotics.

On the other hand, HF-ST + A mice showed the highest content of 2-decenal, 10-undecenoic acid (UDA), cyclohexanol, and fructose as well as the lowest levels of oleic acid and cholesterol compared to HF, HF-ST, and HF-ST + I groups; hence, these metabolites might be used as biomarkers for agavins prebiotic. While, HF-ST + I mice had an increment of three methyl esters: linoleic, oleic, and stearic

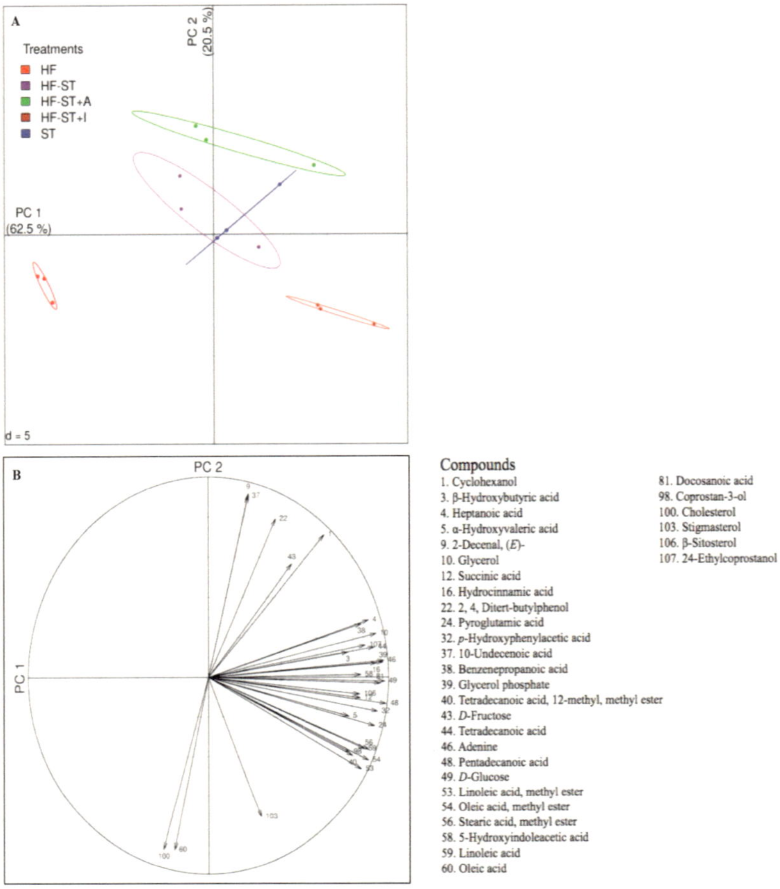

Figure 1.
Metabolites enriched or depleted in the feces of overweight mice after a diet shift and prebiotic supplementation. (A) PCA and (B) loading plot of the two first PCs. Overweight mice (HF) after a diet change (HF-ST) and agavins (HF-ST + A) or inulins (HF-ST + I) supplementation. ST was a healthy mice group.

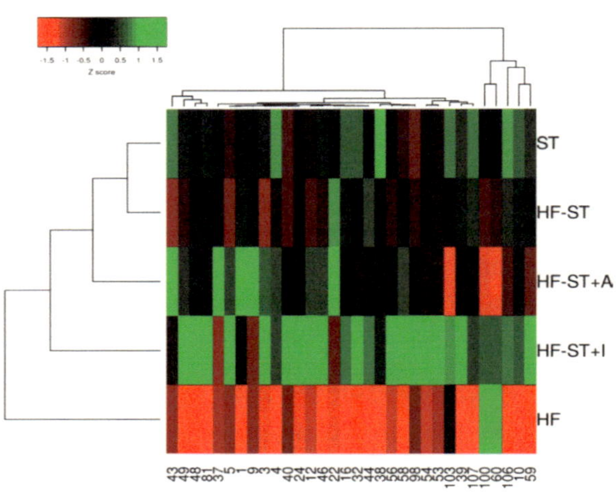

Figure 2.
Heatmap of differential metabolites found in the feces of overweight mice after a diet shift and prebiotic supplementation. HF, overweight mice; HF-ST, overweight mice after a diet shift; HF-ST + A, overweight mice that were switched to standard diet plus agavins supplement; HF-ST + I, overweight mice that were switched to standard diet plus inulins supplement. ST was a healthy mice group.

acid in relation to the other mice groups. These results evidence a clear difference in the fecal metabolites profiles between agavins and inulins, which was associated to their structural differences, such as the presence of fructose observed exclusively in HF-ST + A group. Agavins structure presents at least four terminal fructose units, so that some gut bacteria might start breaking down this prebiotic releasing fructose and agavins of smaller degree of polymerization [30]. In addition, 2-decenal and UDA were the metabolites mostly enriched with agavins intake. 2-decenal has a broad antimicrobial spectrum against pathogenic bacteria [31], while UDA is a neuroprotectant compound [32, 33] used as a nutritional supplement for maintaining a healthy balance of gut microbiota [34]. Besides, UDA also might be acting through GPR84 (a newly described medium chain fatty acid receptor) associated to immunological responses [35], and this mechanism could also contribute to improve the host health. Whereas in comparison to agavins, inulins led to a significant increase of methyl esters and sterols, coinciding with previous studies which proposed this event as a mechanism to improve the lipid metabolism of host [36, 37].

3.2 Organic acid profiles

A total of 21 organic acids were identified, including SCFA, hydroxy-SCFA, dicarboxylic acids, and aromatic carboxylic acids (**Table 3**). PCA analysis of organic acids showed the HF group in a separate cluster, while standard diets (HF-ST and ST) displayed an overlap due to presence of similar metabolites in both groups (**Figure 3A**). In addition, PCA and the loading plot evidenced that HF-ST + A and HF-ST + I groups had a more similar organic acids profiles among them in relation to the other diets (**Figure 3B**).

Moreover, hierarchical clustering analysis evidenced that HF group had the lowest content of all organic acids, while HF-ST + A mice exhibited the highest content of the majority of them; therefore, this group is located at one end of the heatmap (**Figure 4**). Noticeably, HF-ST + I group showed various metabolites with a similar content as HF group; for instance the following acids: 2 methyl butanoic, lactic, and hexanoic (**Table 3**); hence, HF-ST + I appears very close to HF group in the heatmap (**Figure 4**). Once again, HF-ST and ST are the closes groups because they presented similar levels of most analyzed organic acids.

Interestingly, and despite that the hierarchical analyses locate fructans diets very distant from each other (in addition to the great structural differences between agavins and inulins), HF-ST + A and HF-ST + I groups showed an increment of the following acids: succinic, BHB, α-hydroxyisovaleric, and α-hydroxyglutaric. Stunningly, succinic acid is involved in glucose homeostasis [38], while BHB has a neuroprotective effect in mice [39] as well as may inhibit the release of fatty acids from adipose tissue by the hydroxy-carboxylic acid receptor 2 (HCA_2) [40]. Then, these metabolites might be employed in general as biomarkers of prebiotic supplementation.

On the other hand, only agavins supplementation led to a notably enrichment of four acids: nicotinic, 3-phenyllactic, 5-hydroxyvaleric, and lactic; therefore, these metabolites could be used as specifics biomarkers for this prebiotic. It is known that nicotinic acid also stimulate the HCA_2, thereby decreasing plasma lipids and protecting against atherosclerotic disorders [41], and this event could be contributing to improve lipid metabolism of overweight mice that we previously reported [10]. Whereas 3-phenyllactic (the metabolite with the highest increment after agavins consumption) is synthesized by *Lactobacillus* strains and exerts direct antipathogenic activities against bacteria, viruses, and fungi [42].

Surprisingly, we did not find any differential abundance of any organic acid with inulins intake.

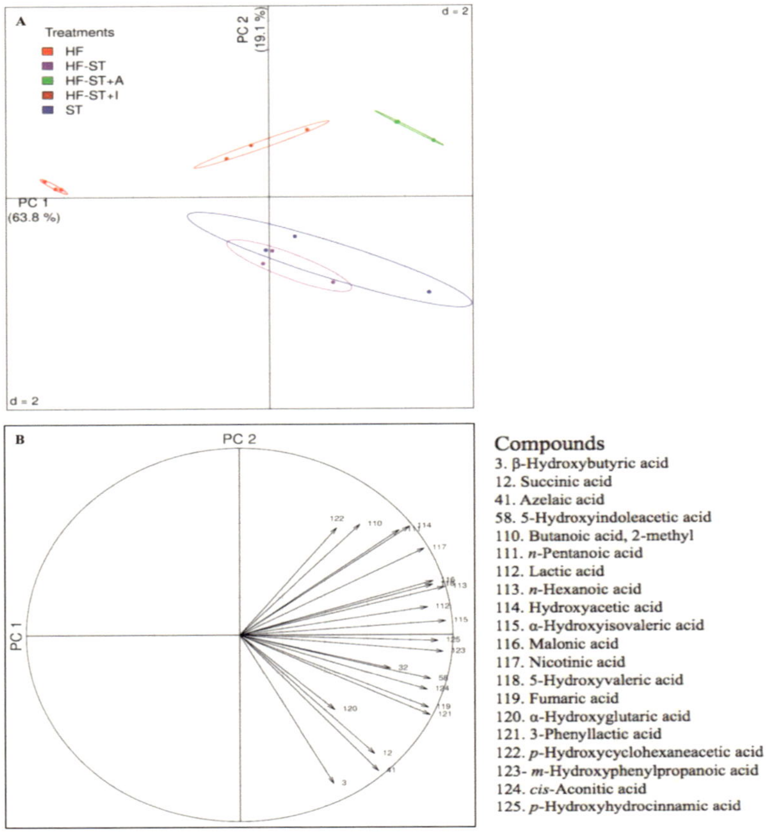

Figure 3.

Organic acids detected in the feces of overweight mice after a diet shift and prebiotic supplementation. (A) PCA and (B) loading plot of the two first PCs. Overweight mice (HF) after a diet change (HF-ST) and agavins (HF-ST + A) or inulins (HF-ST + I) supplementation. ST was a healthy mice group.

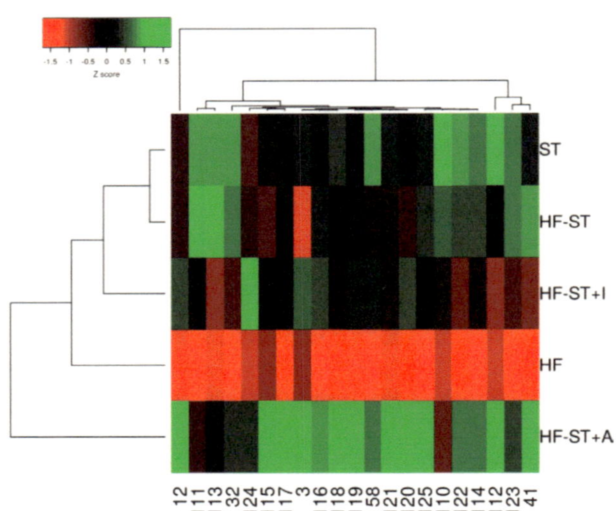

Figure 4.

Hierarchically clustered heatmap of organic acids found in the feces of overweight mice after a diet shift and prebiotic supplementation. HF, overweight mice; HF-ST, overweight mice after a diet shift; HF-ST + A, overweight mice that were switched to standard diet plus agavins supplement; HF-ST + I, overweight mice that were switched to standard diet plus inulins supplement. ST was a healthy mice group.

ID	RT	Compound	Family	tFold change			
				HF	HF-ST	HF-ST + A	HF-ST + I
110	11.015	Butanoic acid, 2-methyl	SCFA	−0.46	−0.15	−0.38	−0.32
111	12.382	*n*-Pentanoic acid	SCFA	−0.71	0.07	−0.37	−0.23
112	14.666	Lactic acid	αOH-SCFA	−1.00	−0.41	0.05	−0.89
113	14.826	*n*-Hexanoic acid	SCFA	−0.66	0.12	−0.27	−0.54
114	15.028	Hydroxyacetic acid	αOH-SCFA	−0.89	−0.08	0.00	−0.39
3	17.177	β-Hydroxybutyric acid	βOH-SCFA	−0.74	−0.98	0.62	0.23
115	17.312	α-Hydroxyisovaleric acid	αOH-SCFA	−0.86	−0.44	2.26	0.11
116	18.248	Malonic acid	DiCAc	−0.53	0.08	0.29	0.18
117	20.133	Nicotinic acid	PyrCAc	−0.80	−0.18	0.56	−0.06
12	20.773	Succinic acid	DiCAc	−0.88	0.01	2.01	0.96
118	21.186	5-Hydroxyvaleric acid	ωOH-SCFA	−1.00	−0.18	0.51	−0.40
119	21.483	Fumaric acid	DiCAc	−1.00	0.29	1.05	0.38
120	26.404	α-Hydroxyglutaric acid	αOH-SCFA	−1.00	−0.24	0.88	0.39
121	26.483	3-Phenyllactic acid	αOH-Ar-SCFA	−0.80	−0.32	0.66	−0.32
122	26.993	*p*-Hydroxycyclohexaneacetic acid	OH-Cy-SCFA	−0.80	−0.15	−0.07	−0.53
32	27.406	*p*-Hydroxyphenylacetic acid	OH-Ar-SCFA	−0.98	−0.02	−0.17	−0.55
123	28.745	*m*-Hydroxyphenylpropanoic acid	PhePr	−0.95	−0.02	−0.10	−0.51
124	29.137	*cis*-Aconitic acid	TCA	−1.00	0.06	1.62	3.48
125	29.186	*p*-Hydroxyhydrocinnamic acid	PhePr	−0.62	0.04	0.18	−0.09
41	29.634	Azelaic acid	DiCAc	−0.69	0.20	0.25	−0.41
58	29.2929	5-Hydroxyindoleacetic acid	IndolAc	−0.90	−0.40	−0.03	−0.22

Fold change value was calculated by comparison with the healthy mice fed with a standard diet (ST). HF, overweight mice; HF-ST, overweight mice that were switched to a standard diet; HF-ST + A, overweight mice changed to standard diet plus agavins; HF-ST + I, overweight mice changed to standard diet plus inulins. All the metabolites listed here have significant difference at least in one treatment p < 0.5. ID numbers correspond with those of ***Table 1****. SCFA, short-chain fatty acid; αOH, alfa-hydroxy; βOH, beta-hydroxy; DiCAc, dicarboxylic acid; PyrCAc, pyridine carboxylic acid; ωOH, omega-hydroxy; Cy, cyclic; Ar, aromatic; PhePr, phenylpropanoid; TCA, tricarboxylic acid; IndolAc, indolic acid.*

Table 3.
Fold-change of differential organic acids detected in the feces of overweight mice after a diet switch and prebiotic supplementation.

4. Conclusions

Microbial metabolites found in agavins group exhibited greater similarity to healthy mice, plus enrichment of specific metabolites (biomarkers) such as 2-decenal, UDA, cyclohexanol, fructose as well as some organic acids that undoubtedly are playing a very important role on overweight mice health. For instance, 2-decenal possess antimicrobial properties; UDA is a neuroprotectant compound; nicotinic acid can decrease plasma lipids levels; while 3-phenyllatic acid shown antipathogenic activities versus bacteria, viruses and fungi. Nevertheless, further studies are needed to clarify the underlying mechanisms by which metabolites derived from agavins fermentation induce a beneficial effect on health of host. Finally, these findings open new and exciting opportunities to explore new biomarkers with applicability on prevention, therapy, or treatment of overweight people.

Acknowledgements

We deeply appreciate Inulina y Miel de Agave, S.A. de C.V. for its constant support.

Conflict of interest

All authors report no financial interests or potential conflicts of interest.

Nomenclature

HF	overweight mice
HF-ST	overweight mice that were switched to a standard diet
HF-ST + A	overweight mice changed to standard diet plus agavins
HF-ST + I	overweight mice changed to standard diet plus inulins
ST	healthy mice
SCFA	short-chain fatty acids
GC/MS	gas chromatography/mass spectrometry
BHB	β-hydroxybutyric acid
UDA	10-undecenoic acid
PCA	principal component analysis

Author details

Alicia Huazano-García, Horacio Claudio Morales-Torres, Juan Vázquez-Martínez and Mercedes G. López*
Department of Biotechnology and Biochemistry, Center of Research and Advanced Studies of the National Polytechnic Institute, Irapuato, Mexico

*Address all correspondence to: mercedes.lopez@cinvestav.mx

References

[1] Flint HJ, Duncan SH, Scott KP, Louis P. Links between diet, gut microbiota composition and gut metabolism. The Proceedings of the Nutrition Society. 2015;**74**:13-22. DOI: 10.1017/s0029665114001463

[2] Turnbaugh PJ. Microbes and diet-induced obesity: Fast, cheap, and out of control. Cell Host and Microbe. 2017;**21**:278-281. DOI: 10.1016/j.chom.2017.02.021

[3] Ley RE, Turnbaugh PJ, Klein S, Gordon JI. Microbial ecology: Human gut microbes associated with obesity. Nature. 2006;**444**:1022-1023. DOI: 10.1038/nature4441022a

[4] Turnbaugh PJ, Bäckhed F, Fulton L, Gordon JI. Diet-induced obesity is linked to marked but reversible alterations in the mouse distal gut microbiome. Cell Host and Microbe. 2008;**3**:213-223. DOI: 10.1016/j.chom.2008.02.015

[5] Mazidi M, Rezaie P, Kengne AP, Mobarhan MG, Ferns GA. Gut microbiome and metabolic syndrome. Diabetes and Metabolic Syndrome. 2016;**10**:S150-S157. DOI: 10.1016/j.dsx.2016.01.024

[6] Dahiya DK, Puniya M, Shandilya UK, Dhewa T, Kumar N, Kumar S, et al. Gut microbiota modulation and its relationship with obesity using prebiotic fibers and probiotics: A review. Frontiers in Microbiology. 2017;**8**:563. DOI: 10.3389/fmicb.2017.00563

[7] He M, Shi B. Gut microbiota as a potential target of metabolic syndrome: The role of probiotics and prebiotics. Cell and Bioscience. 2017;**7**:54. DOI: 10.1186/s13578-017-0183-1

[8] Urias-Silvas JE, Cani PD, Delmée E, Neyrinck A, López MG, Delzenne NM. Physiological effects of dietary fructans extracted from *Agave tequilana* Gto. And *Dasylirion* spp. The British Journal of Nutrition. 2008;**99**:254-261. DOI: 10.1017/S0007114507795338

[9] Santiago-García PA, López MG. Agavins from *Agave angustifolia* and *Agave potatorum* affect food intake, body weight gain and satiety-related hormones (GLP-1 and ghrelin) in mice. Food and Function. 2014;**5**:3311-3319. DOI: 10.1039/c4fo00561a

[10] Huazano-García A, López MG. Agavins reverse the metabolic disorders in overweight mice through the increment of short chain fatty acids and hormones. Food and Function. 2015;**6**:3720-3727. DOI: 10.1039/c5fo00830a

[11] López MG, Mancilla-Margalli NA, Mendoza-Diaz G. Molecular structures of fructans from *Agave tequilana* Weber var. Azul. Journal of Agricultural and Food Chemistry. 2003;**51**:7835-7840. DOI: 10.1021/jf030383v

[12] Mancilla-Margalli NA, López MG. Water-soluble carbohydrates and fructan structure patterns from agave and Dasylirion species. Journal of Agricultural and Food Chemistry. 2006;**54**:7832-7839. DOI: 10.1021/jf060354v

[13] Gibson GR, Scott KP, Rastall RA, Tuohy KM, Hotchkiss A, Dubert-Ferrandon A, et al. Dietary prebiotics: Current status and new definition. Food Science and Technology Bulletin. 2010;**7**:1-19. DOI: 10.1616/1476-2137.15880

[14] Huazano-García A, Shin H, López MG. Modulation of gut microbiota of overweight mice by agavins and their association with body weight loss. Nutrients. 2017;**9**:E821. DOI: 10.3390/nu9090821

[15] Flint HJ, Scott KP, Duncan SH, Louis P, Forano E. Microbial degradation of complex carbohydrates in the gut. Gut Microbes. 2012;**3**:289-306. DOI: 10.4161/gmic.19897

[16] Cani PD, Van Hul M, Lefort C, Depommier C, Rastelli M, Everard A. Microbial regulation of organismal energy homeostasis. Nature Metabolism. 2019;**1**:34-46. DOI: 10.1038/s42255-018-0017-4

[17] García-Curbelo Y, Bocourt R, Savón LL, García-Vieyra MI, López MG. Prebiotic effect of *Agave fourcroydes* fructans: An animal model. Food and Function. 2015;**6**:3177-3182. DOI: 10.1039/C5FO00653H

[18] Ramnani P, Costabile A, Bustillo AGR, Gibson GR. A randomised, double-blind, cross-over study investigating the prebiotic effect of agave fructans in healthy humans subject. Journal of Nutritional Science. 2015;**4**:e10. DOI: 10.1017/JNS.2014.68

[19] Psichas A, Sleeth ML, Murphy KG, Brooks L, Bewick GA, Hanyaloglu AC, et al. The short chain fatty acid propionate stimulates GLP-1 and PYY secretion via free fatty acid receptor 2 in rodents. International Journal of Obesity. 2015;**39**:424-429. DOI: 10.1038/ijo.2014.153

[20] Tolhurst G, Heffron H, Lam YS, Parker HE, Habib AM, Diakogiannaki E, et al. Short-chain fatty acids stimulate glucagon-like peptide-1 secretion via the G-protein-coupled receptor FFAR2. Diabetes. 2012;**61**:364-371. DOI: 10.2337/db11-1019

[21] Lin HV, Frasseto A, Kowalik EJ Jr, Nawrocki AR, Lu MM, Kosinski JR, et al. Butyrate and propionate protect against diet-induced obesity and regulate gut hormones via free fatty acids receptor 3-independent mechanisms. PLoS One. 2012;**7**:e35240. DOI: 10.1371/journal.pone.0035240

[22] De Vadder F, Kovatcheva-Datchary P, Goncalves D, Vinera J, Zitoun C, Duchampt A, et al. Microbiota-generated metabolites promote metabolic benefits via gut-brain neural circuits. Cell. 2014;**156**:84-96. DOI: 10.1016/j.cell.2013.12.016

[23] Singh V, Chassaing B, Zhang L, San Yeoh B, Xiao X, Baker MT, et al. Microbiota-dependent hepatic lipogenesis mediated by stearoyl CoA desaturase 1 (SCD1) promotes metabolic syndrome in TLR5-deficient mice. Cell Metabolism. 2015;**22**:983-996. DOI: 10.1016/j.cmet.2015.09.028

[24] Wishart DS. Emerging application of metabolomics in drug discovery and precision medicine. Nature Reviews. Drug Discovery. 2016;**15**:473-484. DOI: 10.1038/nrd.2016.32

[25] Hamer HM, De Preter V, Windey K, Verbeke K. Functional analysis of colonic bacterial metabolism: Relevant to health? American Journal of Physiology. Gastrointestinal and Liver Physiology. 2012;**302**:G1-G9. DOI: 10.1152/ajpgi.00048.2011

[26] Eneroth P, Hellstroem K, Ryhage R. Identification and quantification of neutral fecal steroids by gas-liquid chromatography and mass spectrometry: Studies of human excretion during two dietary regimens. Journal of Lipid Research. 1964;**5**:245-262

[27] Gao X, Pujos-Guillot E, Sébédio JL. Development of a quantitative metabolomic approach to study clinical human fecal water metabolome based on trimethylsilylation derivatization and GC/MS analysis. Analytical Chemistry. 2010;**82**:6447-6456. DOI: 10.1021/ac1006552

[28] García-Villalba R, Giménez-Bastida JA, García-Conesa MT, Tomás-Barberán FA, Carlos Espín J, Larrosa M.

Alternative method for gas chromatography-mass spectrometry analysis of short-chain fatty acids in fecal samples. Journal of Separation Science. 2012;**35**:1906-1913. DOI: 10.1002/jssc.201101121

[29] Matysik S, Le Roy CI, Liebisch G, Claus SP. Metabolomics of fecal samples: A practical consideration. Trends in Food Science and Technology. 2016;**57**:244-255. DOI: 10.1016/j.tifs.2016.05.011

[30] Huazano-García A, López MG. Enzymatic hydrolysis of agavins to generate branched fructooligosaccharides (a-FOS). Applied Biochemistry and Biotechnology. 2018;**184**:25-34. DOI: 10.1007/s12010-017-2526-0

[31] Bisignano G, Laganà MG, Trombetta D, Arena S, Nostro A, Uccella N, et al. *In vitro* antibacterial activity of some aliphatic aldehydes from *Olea europaea* L. FEMS Microbiology Letters. 2001;**198**:9-13. DOI: 10.1111/j.1574-6968.2001.tb10611.x

[32] Jantas D, Piotrowski M, Lason W. An involvement of PI3-K/Akt activation and inhibition of AIF translocation in neuroprotective effects of undecylenic acid (UDA) against pro-apoptotic factors-induced cell death in human neuroblastoma SH-SY5Y cells. Journal of Cellular Biochemistry. 2015;**116**:2882-2895. DOI: 10.1002/jcb.25236

[33] Lee E, Eom JE, Kim HL, Kang DH, Jun KY, Jung DS, et al. Neuroprotective effect of undecylenic acid extrated from *Ricinus communis* L. through inhibition of μ-calpain. European Journal of Pharmaceutical Sciences. 2012;**46**:17-25. DOI: 10.1016/j.ejps.2012.01.015

[34] Brayden DJ, Walsh E. Efficacious intestinal permeation enhancement induced by the sodium salt of 10-undecylenic acid, a medium chain fatty acid derivative. The AAPS Journal. 2014;**16**:1064-1076. DOI: 10.1208/s12248-014-9634-3

[35] Alvarez-Curto E, Milligan G. Metabolism meets immunity: The role of free fatty acid receptors in the immune system. Biochemical Pharmacology. 2016;**114**:3-13. DOI: 10.1016/j.bcp.2016.03.017

[36] Yang J, Zhang S, Henning SM, Lee R, Hsu M, Grojean E, et al. Cholesterol-lowering effects on dietary pomegranate extract and inulin in mice fed an obesogenic diet. The Journal of Nutritional Biochemistry. 2018;**52**:62-69. DOI: 10.1016/j.jnutbio.2017.10.003

[37] Catry E, Bindels LB, Tailleus A, Lestavel S, Neyrinck AM, Goossens JF, et al. Targeting the gut microbiota with inulin-type fructans: Preclinical demonstration of a novel approach in the management of endothelial dysfunction. Gut. 2018;**67**:271-283. DOI: 10.1136/gutjnl-2016-313316

[38] Hoque M, Ali S, Hoda M. Current status of G-protein coupled receptors as potential targets against type 2 diabetes mellitus. International Journal of Biological Macromolecules. 2018;**118**:2237-2244. DOI: 10.1016/j.ijbiomac.2018.07.091

[39] Rahman M, Muhammad S, Khan MA, Chen H, Ridder DA, Müller-Fielitz H, et al. The β-hydroxybutyrate receptor HCA2 activates a neuroprotective subset of macrophages. Nature Communications. 2014;**5**:3944. DOI: 10.1038/ncomms4944

[40] Taggart AK, Kero J, Gan X, Cai TQ, Cheng K, Ippolito M, et al. (D)-beta-hydroxybutyrate inhibits adipocyte lipolysis via the nicotinic acid receptor PUMA-G. The Journal of Biological Chemistry. 2005;**280**:26649-26652. DOI: 10.1074/jbc.c500213200

[41] Lukasova M, Hanson J, Tunaru S, Offermanns S. Nicotinic acid (niacin): New lipid-independent mechanism of action and therapeutic potentials. Trends in Pharmacological Sciences. 2011;**32**:700-707. DOI: 10.1016/j.tips.2011.08.002

[42] Zhang Z, Lv J, Pan L, Zhang Y. Roles and applications of probiotic lactobacillus strains. Applied Microbiology and Biotechnology. 2018;**102**:8135-8143. DOI: 10.1007/s00253-018-9217-9

Section 2

Metabolomics: New Insights into Medicine

Chapter 3

Metabolomic Discovery of Microbiota Dysfunction as the Cause of Pathology

Natalia V. Beloborodova, Andrey V. Grechko and Andrey Yu Olenin

Abstract

In the twenty-first century, metabolomics allowed evaluating the profile of metabolites of various classes of compounds in the human body. The most important achievement of the metabolic approach is to obtain evidence of the intersection of human biochemical pathways and its microbiota. The effect of certain microbial metabolites on the work of key enzymes involved in the biotransformation of amino acids and other substances becomes more important in patients at risk of developing neurological and mental disorders and also contributes to the development of life-threatening conditions up to multiple organ failure after operations, injuries, and serious diseases. The authors of this chapter call the microbiota an "invisible organ," emphasizing its functional significance, and not just taxonomy, as previously thought. This chapter will discuss the mutually beneficial integration of the metabolome/microbiome in the body of healthy people and will focus on the effects of microbiota dysfunction.

Keywords: homeostasis, microbiota, "invisible organ," bacterial metabolites

1. Introduction

Homeostasis is key for the normal performance of a human body. Many parameters are constantly maintained in fairly narrow vital ranges, such as temperature, acidity in the intracellular and intercellular spaces, the electrolyte concentrations, hormones, vitamins, etc. The traditional view is that the body itself is able to maintain the constancy of its internal environment due to a complex system of feedback (**Figure 1**). Each organ helps to maintain homeostasis, ensuring its specific function. It acts as a backward force that returns the system to equilibrium in the event of deviations from the normal state. Along with other organs, the microbiota plays an important role in maintaining homeostasis, despite being an "invisible organ."

By the way, in terms of weight, the microbiota should be attributed to the largest organ that can be compared only with the brain or liver: this can be easily ascertained using simple calculations based on known facts about the weight of human organs relative to the body weight of an adult (**Figure 2**). The human microbiota, which is a community of gut microorganisms, can be considered as an independent organ with many functions.

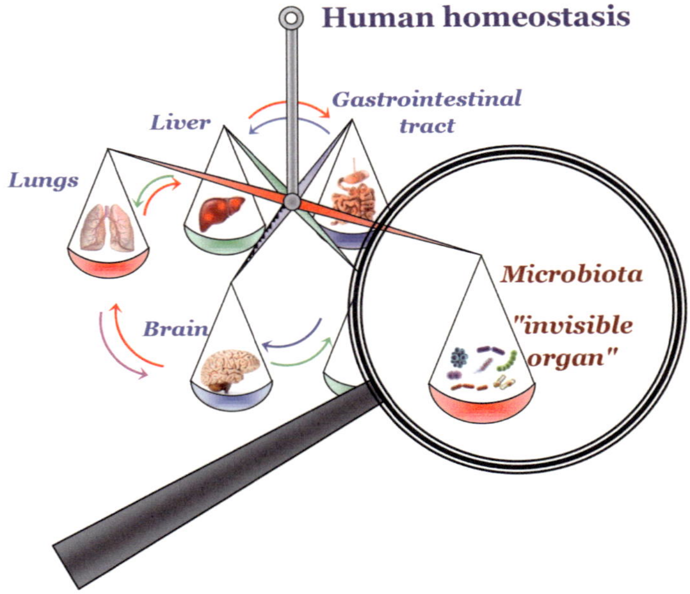

Figure 1.
Diagram of the interaction of organs that support the state of homeostasis.

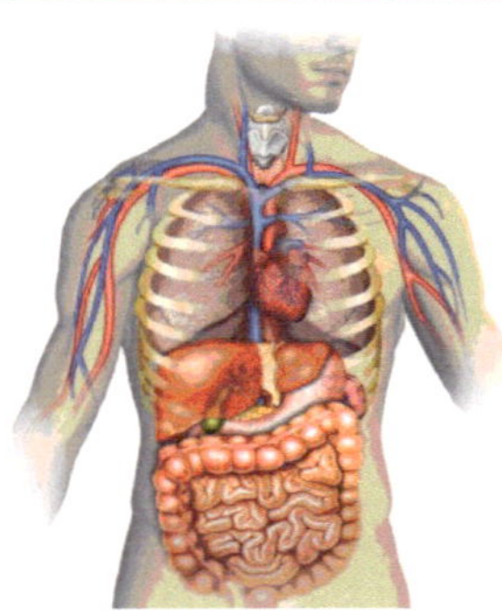	*Organ*	*Weight, g*	*Part of body mass, %*
	• *kidney*	*250*	*0,35*
	• *heart*	*300*	*0,4*
	• *lungs*	*1000*	*1,4*
	• *brain*	*1500*	*2,1*
	• *liver*	*1700*	*2,4*
	• ***microbiota***	***1000 - 2000***	***1-3***

Figure 2.
Microbiota as a big but "invisible organ," % of body mass compared to other vital organs in an adult weighing 70 kg.

In the twenty-first century, a new insight on the processes occurring in the human body in health and disease on the basis of the new knowledge of the microbiota is formed. Detection and identification of the trillions of bacteria that form the microbiota of healthy and sick people are made possible by the use of modern technologies, for example, sequencing of the 16S rRNA gene.

Metabolite-based approaches (or metabolomics) to the study of the human microbiota are more significant progress in biology and medicine, searching for answers to the question "What are the chemical and pathophysiological results of the metabolic activity of the microbiota?" Today many research teams are searching for answers to this question [1].

The host organism is a habitat for the microbiota, so maintaining homeostasis is vital for the survival of hundreds of bacterial species. The microbiota seeks to restore homeostasis in the case of minor metabolic disorders that are not systemic in nature, and it has a huge amount of possibilities for this. If changes in the vital functions of the body are serious, a new quality (pathology) is formed, the microbiota is also radically rebuilt: this is manifested not only in changes in the species composition of bacteria (taxonomy) but also in metabolic processes. Other non-normal products of microbial metabolism from the intestines enter the systemic circulation, and they can interfere with the endogenous metabolic pathways. When the microbiota works against the host, it is manifested by diseases, even death (sepsis).

The medical community has not yet formed an understanding of the role of the microbiota as a separate organ. A search query ("microbiota as an organ") or ("microbiome as an organ") in specialized databases, such as the Web of Science, Scopus, and Pubmed, gives a negative result. At the same time, a number of review articles are actually present which describe in detail the physiology and biochemistry of the close interaction of the intestinal microbiota with the host organism, in which there are many qualities and attributes of the organ.

This chapter formulates ideas about the microbiota as an organ, which has become possible due to the results of studies with metabolomic equipment of recent years. The material presented in this chapter relies primarily on articles published after 2010. Specialists working in both fundamental and clinical medicine are undoubtedly interested in the growing information about the role of microbiota in maintaining homeostasis, as well as the participation of microorganisms of the human body in the metabolic pathways, which are directly related to the development of various pathologies.

2. Microbiota in a healthy body

Food intake, its conversion, and excretion of waste products are material sources for the normal functioning of a human body. The aim of nutrition from a biochemical viewpoint is to maintain the body's critical parameters in narrowly defined value rates. The concept of a "living healthy organism" consists precisely in the ability to resist change and maintain the constancy of the composition and properties of its internal environment. The basis of digestion is a fairly universal mechanism, which includes splitting of the main components, such as carbohydrates (including polysaccharides), fats, biopolymers (proteins, macromolecules based on nucleotide sequences), etc., to individual low-molecular substances and then to the synthesis of low- and high-molecular weight compounds, which are the material basis for cells and organs as well as the energy source for biochemical reactions. Interest to low-molecular weight compounds has grown particularly in recent years. The Human Metabolome Database (HMDB) was created and is constantly updated by the international researcher group. Now it contains information on more than 100,000 individual low-molecular compounds (metabolites), constituting about 25,000 pathways of metabolism [2].

Food digestion is one of the main complex processes that form homeostasis. Transformation of the matter occurs throughout the gastrointestinal tract. Food undergoes ever-deeper processing as you move through it. Enzymes directly involved in this can potentially have endogenous and exogenous origin. The endogenous pathway is carried out with the participation of its own secrets produced by the body with the participation of organs that promote digestion and the excretion of waste products. The complex of biochemical reactions that coincide with the active participation of the microbiota, consisting of hundreds, sometimes reaching up to several thousand species, is presented as an alternative to it. In the literature there is no single point of view about the density of microorganism colonization of the human digestive system. According to [3], the relative content of microorganisms (cells/mL) in different parts of the gastrointestinal tract is duodenum, 10^1–10^3; jejunum and ileum, 10^4–10^7; cecum, 10^8; and large intestine, 10^{11}–10^{12}. A large number of publications give the relative content of microorganisms in the range of 10^2–10^{13}, while the maximum values are recorded in the cecum and transverse colon.

The specificity of food digestion is due to the variety of enzymes capable of carrying out similar biochemical transformations, if not entirely, then at least of its many components, due to intestinal microbiota. The synthesis of specific proteins, including enzymes, is due to the presence of various nucleotide DNA sequences. The diversity of these sequences in a complex system consisting of hundreds, or even thousands, of individual species of microorganisms is significantly higher than that of human. The lifetime of a particular microorganism, depending on the immune response of the host organism, correlates with the function that promotes or interferes with its vital activity. The production of specific microorganism killer proteins is not observed in the case of symbiosis. Processes of synthesis of interleukins and phagocytosis are immediately activated in the alternative situation [4]. A big array of metagenomic studies of human intestinal microbiota collected in recent years in various information repositories, such as the National Center for Biotechnology Information (NCBI).

The role of microbiota is quite significant already at the stage of primary processing of nutrients. For example, in [5], the fact is given that only bacteroids of the *Bacteroides thetaiotaomicron* contain nucleotide sequences for the synthesis of 260 glycosidic hydrolases, while the entire human genome is capable of producing only 17 such enzymes, and 9 of them are not fully characterized. The author of the review [6] provides several specific metabolic pathways associated with intestinal microbiota. These include (i) cleavage of polysaccharides to monomers, followed by processing into short-chain fatty acids; (ii) depolymerization of proteins to amino acids, with further conversion of some of them (glycine, lysine, arginine, leucine, isoleucine, and valine) to nitrogen-containing heterocyclic compounds, for example, substituted indoles; (iii) neutralization and detoxification of arene-containing components from the external environment; and (iv) biotransformation of fats and bile acids and their inclusion in biochemical processes that promote energy cells, for example, in the Krebs cycle.

The species composition of the microbiota is specific for each person and depends on many factors, such as age, diet, use of antibiotics, etc. We can talk about two components of the microbiota—obligate or transient. A self-organizing ecosystem with the dominance of some species of microorganisms and the oppression of others arises in a normally functioning organism. The classification and systematization of information on the species and genetic diversity of the microbiota of the human body were carried out independently by two scientific communities in the United States and the European Union, which resulted in the appearance of two databases: Human Microbiome Project (HMP) [7] and Metagenomics of the Human Intestinal Tract (MetaHIT) [8].

Extensive information on the composition of the intestinal microbiota of a healthy person is contained in the literature. These studies indicate the dominance of several genera of strict anaerobes, and the main ones are *Bacteroides*, *Prevotella*, *Eubacterium*, *Ruminococcus*, *Clostridium*, *Lactobacillus*, and *Bifidobacterium*. The data on the microbial community of the gastrointestinal tract are summarized in detail in the 2018 review [9] and presented in **Table 1**, which reflects the gradual change in the species composition of microorganisms as food progresses and digests.

A huge number of types of microorganisms perform the biochemical functions which we call the "conveyor" of the microbiota [10]. The diversity of species with different biochemical activity provides coordinated work of the microbiota. The final metabolite formation depends on many factors: the quality and quantity of substrate (food components); the function of the stomach, pancreas, liver, and gallbladder; bowel motility; etc. definitely influence the metabolism of microbiota. The normal biotransformation of any of the substrates in the intestinal lumen takes place sequentially [10]. Biochemistry of deep food transformation is in many respects similar to the metabolic characteristic of microorganisms. The main part of

Part of the gastrointestinal tract	The dominant species composition of the microbiota
Oral cavity	*Gemella, Granulicatella, Streptococcus, Prevotella, Veillonella, Porphyromonas, Neisseria, Rothia, Lactobacillus, Fusobacterium*
Throat, esophagus	*Streptococcus, Prevotella, Actinomyces, Gemella, Rothia, Granulicatella, Haemophilus, Veillonella*
Stomach	*Helicobacter pylori, Veillonella, Lactobacillus*
Small intestine	*Enterococcus, Escherichia coli, Klebsiella, Lactobacillus, Staphylococcus, Streptococcus, Bacteroides fragilis, Clostridium lituseburense, Gammaproteobacterium*
Cecum	*Lactobacillus, Enterococcus, Escherichia coli, Bacteroides, Clostridium leptum, Clostridium coccoides*
Rising gut	*Bacteroides, Lactobacillus, Bifidobacterium*
Colon	*Bacteroides, Clostridium, Desulfomonas, Desulfovibrio*

Table 1.
Differences in the composition of the microbiota throughout the gastrointestinal tract (adapted from [9]).

the individual amino acids that come from food after the cleavage of polypeptides is further spent on the synthesis of its own proteins, which are necessary for the functioning of the body. Residual amino acids can be transformed into other substances of a non-protein nature, performing a number of important functions not related to digestion or the building function.

This trend is most pronounced for aromatic amino acids such as phenylalanine, tyrosine, and tryptophan. The transformations of the phenylalanine-tyrosine pair occurring in the liver are contained in the human metabolome database (**Figure 3**). Phenylalanine and tyrosine are interchangeable in terms of metabolism. Phenylalanine is transformed into tyrosine under the action of a complex compound of Fe^{2+} ions with phenylalanine-4-hydroxylase with the participation of L-erythrotetrahydrobiopretin. Then both amino acids are transformed into 4-hydroxyphenylpyruvic acid, and then, by successive transformations, they are transformed into acetoacetic and fumaric acids—components of the Krebs cycle under the action of the same enzymes with the participation of the same substances [11]. There is no direct conversion of phenylpyruvic acid to 4-hydroxyphenylpyruvic acid in this metabolism scheme.

The pathway of tyrosine processing, namely, its biotransformation in tyramine further into three directions—dopamine, homovanillin, and dopachinone—is important for the normal functioning of human mental activity (**Figure 4**). All biochemical transformations that make up these metabolic pathways occur with the direct action of enzymes. However, enzymes for not all reactions are listed in the HMDB. The label "??" (**Figure 4**) refers to the absence of data on the enzyme. The pathway reactions can be divided into two types: "traditional" and "unusual."

The first type is rather trivial transformations, such as the conversion of an aldehyde to the corresponding carboxylic acid, for example, homovanillin to homovanillic acid. Such transformations are well known in classical organic chemistry. These reactions do not require enzymes; it is enough to have an oxidizing agent, such as molecular oxygen, hydrogen peroxide, reactive oxygen species, etc. The situation is different in the case of the formation of nitrogen-containing heterocycles formed from aromatic amino acids. The information about enzyme in HMDB is not available for the key dopachinone conversion reaction to leukodopachrome. A detailed study of the mechanism of this reaction, contained in [12], shows that nitric oxide (I) takes an active part in it. This fact is complicated only by understanding the base of interactions. Many reactions of

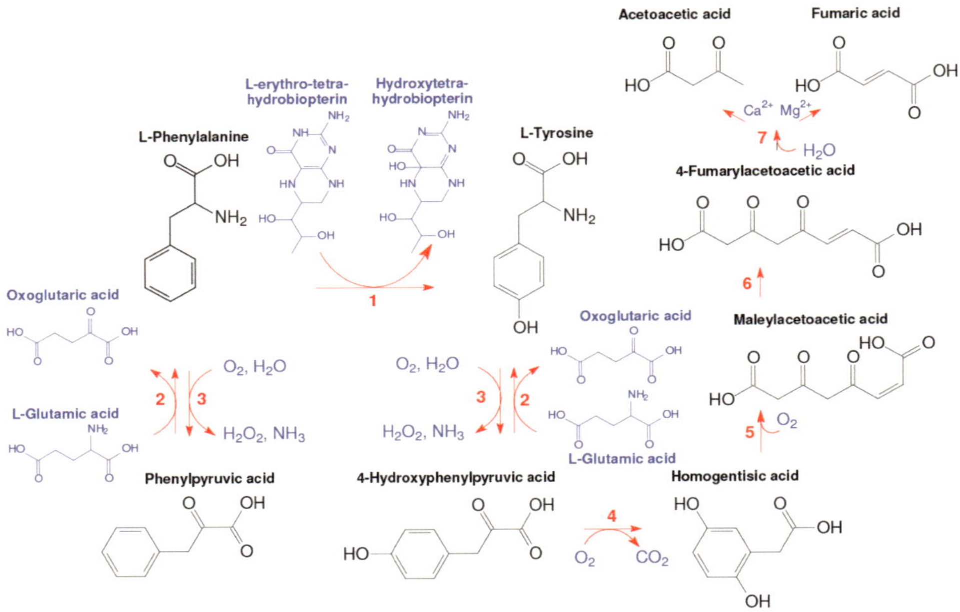

Figure 3.
Normal metabolism of phenylalanine and tyrosine in the liver. Enzymes (coenzymes): (1) phenylalanine-4-hydroxylase (Fe^{2+}); (2) aspartate aminotransferase, cytoplasmic tyrosine aminotransferase; (3) L-amino-acid oxidase (FAD); (4) 4-hydroxyphenylpyruvate dioxygenase (Fe^{3+}); (5) homogentisate 1,2-dioxygenase; (6) maleylacetoacetate isomerase; (7) fumarylacetoacetase (according to the HMDB).

tyrosine metabolism are supported by a complex of copper ions with tyrosinase, well known in the biochemistry of microorganisms and used in biotechnology (see, e.g., [13]).

A significant part of the reactions given in **Figure 4**, with a sufficient degree of confidence, occurs with the participation of enzymes of exogenous (microbiological) origin generated by the microbiota. The formation of metabolites not only with the benzene but also with the indole ring occurs as a result of tyrosine biotransformation, including the participation of microbial enzymes.

Indoles, including those synthesized using human microbiota enzymes, play an important role in metabolism. Such physiologically important substances as serotonin, tryptamine, and derivatives of quinic acid belong to them (**Figure 5**). Many of the compounds involved in the indole metabolism are able to pass through the blood–brain barrier. About 95% of tryptophan enters the brain as a conjugate with kynurenine compounds, whose final metabolic products are kinuric and quinolinic acids, 3-hydroxykynurenine [14].

Reactions associated with the presence of endogenous enzymes and enzymes of microbial origin are in a state of dynamic equilibrium with the normal functioning of biochemical processes in the body. Microbiota metabolism is able to quickly adjust in a direction that helps to maintain homeostasis with moderate deviations (abnormalities with dietary errors, travels with changing time zones, etc.). The dynamic metabolism of the "invisible organ" is provided by the potential of the metabolic pathways, such as catecholamine biosynthesis (**Figure 6**) with participation of numerous species of microorganisms.

Microbiota metabolism can also be seriously affected if the disorders are systemic under the influence of adverse external factors (e.g., massive antimicrobial therapy, severe poisoning, hypoxia, blood loss, etc.). These disorders can manifest themselves clinically by developing a critical state, which often puts the existence of the organism (its life) at risk.

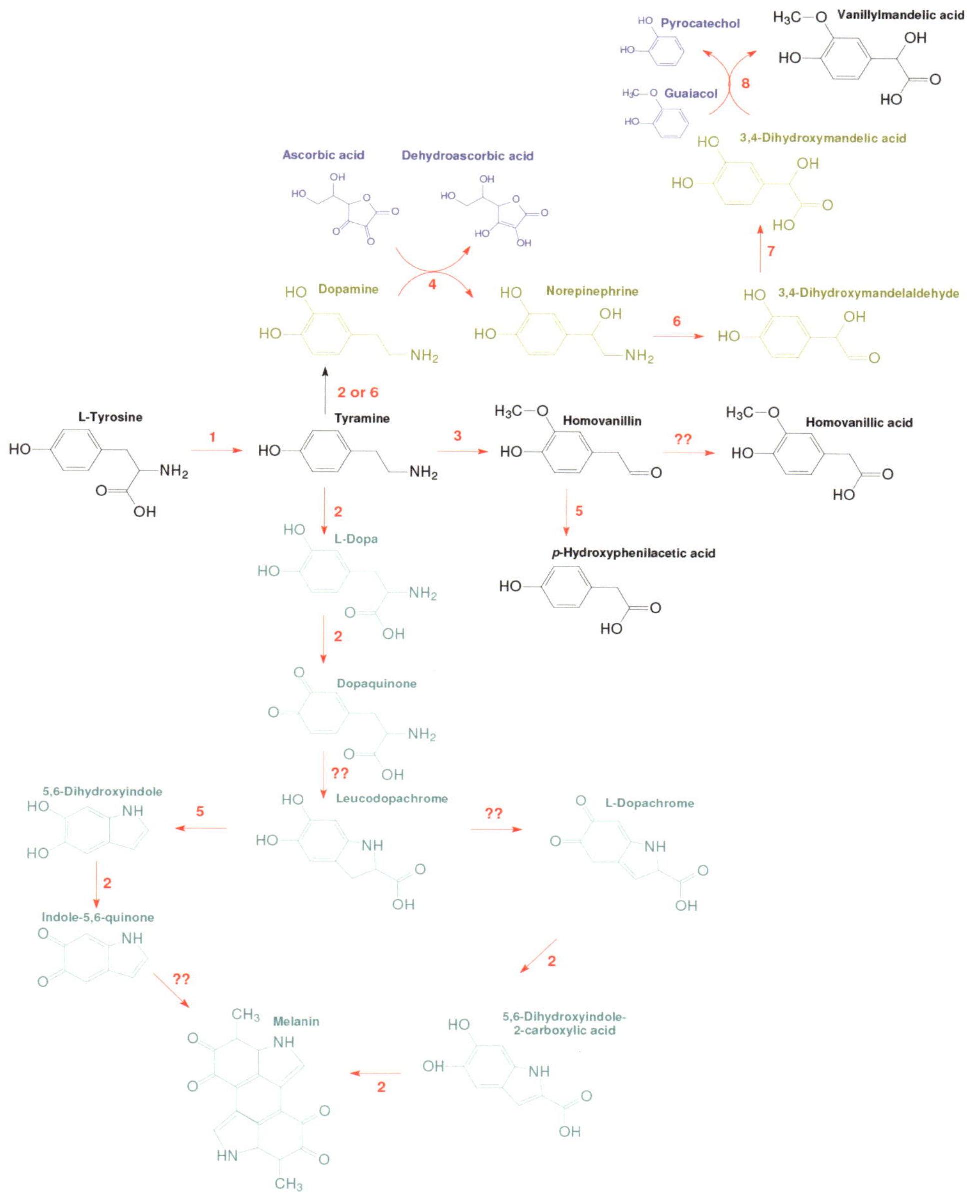

Figure 4.
Tyrosine metabolism. Enzymes (coenzymes): (1) aromatic-L-amino-acid decarboxylase, (Pyridoxal-5′-phosphate); (2) tyrosinase (Cu^{2+}); (3) amiloride-sensitive amine oxidase [copper-containing] (Cu^{2+}, Ca^{2+}, topaquinone); (4) dopamine beta-hydroxylase (Cu^{2+}, pyrroloquinoline, quinone); (5) aldehyde dehydrogenase (dimeric NADP-preferring); (6) amine oxidase [flavin-containing] A (FAD); (7) aldehyde dehydrogenase, (dimeric NADP-preferring); (8) catechol O-methyltransferase (Mg^{2+}) (according to the HMDB).

3. Microbiota dysfunction in pathology

3.1 Diseases of the digestive tract

Disturbances in the normal functioning of the gastrointestinal tract are largely due to changes in the digestion processes associated with the state of the microbiota. As noted above, the microbiota composition depends on the heredity and health of the host, climate, nutrition, bad habits, etc. A system itself is able to return to a state of homeostasis in the case of mild disorders. The microbiota has mechanisms to adapt to the effects of antibacterial substances. Antibiotics are originally the products of bacteria which they use as competitive advantage in the conditions

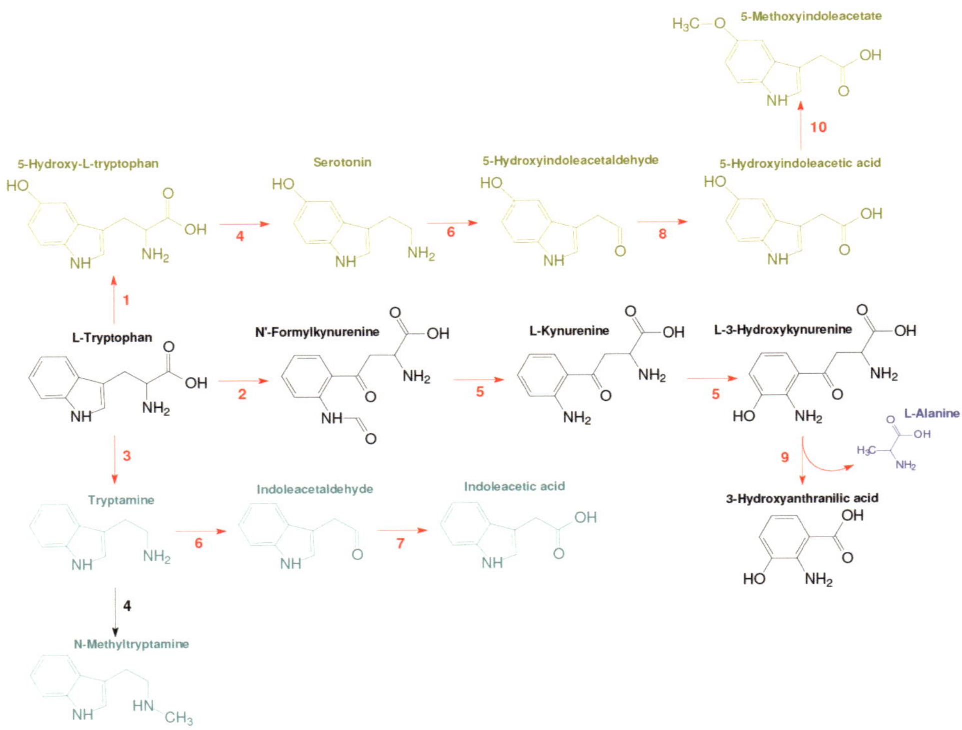

Figure 5.
Simplified scheme of normal tryptophan metabolism. Enzymes (coenzymes): (1) tryptophan 5-hydroxylase (Fe^{2+}); (2) tryptophan 2,3-dioxygenase (heme); (3) aromatic-L-amino-acid decarboxylase (pyridoxal-5'-phosphate); (4) indolethylamine N-methyltransferase; (5) kynurenine formamidase; (6) kynurenine 3-monooxygenase (FAD); (7) aldehyde dehydrogenase, mitochondrial (NAD); (8) aldehyde dehydrogenase, mitochondrial or aldehyde oxidase (FAD, molybdopterin, 2Fe-2S); (9) kynureninase (pyrophosphate); (10) acetylserotonin O-methyltransferase (S-Adenosyl methionine) (according to the HMDB).

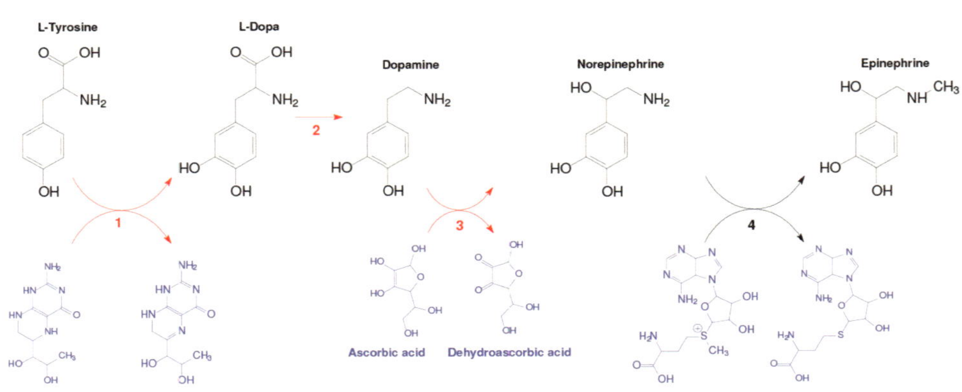

Figure 6.
Catecholamine biosynthesis. Enzymes (coenzymes): (1) tyrosine 3-monooxygenase (Fe^{2+}); (2) aromatic-L-amino-acid decarboxylase (pyridoxal-5'-phosphate); (3) dopamine beta-hydroxylase (Cu^{2+}, pyrroloquinoline, quinone); (4) phenylethanolamine N-methyltransferase) (according to the HMDB).

of nutrient substrate deficiency in their habitat. However, significant changes in species composition may be developed under the influence of broad-spectrum antibacterial drugs, since the massive use of antibiotics (xenobiotics) is violent and anti-biological and can disrupt biochemical processes.

The microbiota is involved in the transformation of xenobiotics and provides a range of reactions including acetylation, deacylation, decarboxylation, dehydroxylation, demethylation, etc. under the influence of low-quality products and synthetic drugs [15]. Modern possibilities of metabolic methods allow an extensive study of another important function of the microbiota—detoxification of the host organism, which maintains its normal state longer in the conditions of retention and self-repairing of the microbiota.

Disorders of the microbial products of short-chain fatty acids SCFA (acetic, propionic, butyric) are most thoroughly studied as a result of the suppression of the normal functioning of anaerobic bacteria. Normally, SCFA requires enterocytes as the main source of energy, respectively; their deficiency contributes to the violation of mucosal trophism, reduction of reparative processes, development of ulcers, and inflammation. Persistent indigestion disorders and chronic gastroenterological diseases are the clinical manifestations of serious changes in the species composition and dysfunction of the microbiota.

Different genera of anaerobic bacteria are called responsible for the production of SCFA. For example, large amount of carbohydrate dissimilation butyrate from dissimilation is associated with some *Clostridia* clusters, other SCFAs, and *Bifidobacterium* spp.

In his review, Nyangale et al. rightly noted that several members of the microbiota have been linked with diseases mainly affecting the gut, lake inflammatory bowel disease, such as ulcerative colitis, Crohn's disease, colorectal cancer, and irritable bowel syndrome, although mechanisms involved are still not yet fully understood [16]. The authors consider the possibilities of metabolite analysis to assess the metabolic activity of the microbiota, to measure volatile and nonvolatile metabolite in biological samples, and to give metabolic pathways the contribution of microbiota to which it is most pronounced. These pathways include also the transformation of glucose and amino acids into SCFA, amino acid, microbial degradation of tyrosine to *p*-hydroxyphenylacetic and *p*-hydroxyphenylbenzoic acids (including bypassing tyramine), and degradation of tryptophan to indolepropionate and indoleacetate (including bypassing tryptamine).

A metabolite composition, determined in the feces, may indicate the composition of microbiota and its changes associated with the use of antibiotics [17]. The use of chemometric approaches in relation to the primary mass spectral data of the samples under study allows one to reliably find the differences between patients with inflamed intestines and the control group. The authors consider that changes in the microbiota phenotype cause this kind of deviations. The ratio of the species composition of microbiota—obligate or transient—significantly affects the metabolite composition that enters the circulatory system from the bowel. Thus, the role of *Bacillus* and *Lactobacillus*, colonizing the epithelium of the gastrointestinal tract, is systematically examined in a review of Ilinskaya et al. [18]. The metabolite composition depends significantly on the activity of their enzymatic systems, even with a relatively low content of such microorganisms in the microbiota.

In such acquired endocrinological diseases as obesity, type 2 diabetes (not related to heredity) can be attributed to pathological conditions due to metabolic disorders involving the microbiota. Microbiota can influence the development of diabetes [19]. Changes in the microbiological composition—dysbacteriosis—caused, for example, by the use of antibiotics, may contribute to an increase in insulin dysfunction, a long-term consequence of which is the development of type 2 diabetes. Due diet may ensure opportune correction of the microbiota and prevent further development of the disease. In a similar study for type 2 diabetes, cited in [20], the authors come to analogous conclusions. The authors agree that function is more important than taxonomy when discussing the role of microbiota in the development of metabolic disorders and diseases of the gastrointestinal tract [21].

In the future, methods of diagnosing gastrointestinal diseases and methods of treatment through the modulation of the microbiota based on information about intermediate metabolites and end products of microbial biodegradation of various compounds can be constructed and developed.

3.2 Microbial metabolites in oncology

Changes in the human body due to microbiota metabolism can affect cells and tissues and contribute to the development of benign and malignant tumors. The biochemistry and physiology of oncological processes is not completely clear, but certain metabolic shifts can be fixed instrumentally for some types of oncological diseases [22, 23]. The successful search for links between the patterns of normal functioning of the microbiota and the biochemistry of carcinogenesis is detailed in recent reviews [24, 25]. This indicates the prospects of such concept and allows us to call the microbiota "a key orchestrator of cancer therapy."

Most of the data on the correlation between a microbiota and cancer tumors is in the gastroenterology [26–31]. Such intestinal microorganisms as *Fusobacterium nucleatum*, *Streptococcus gallolyticus*, *Bacteroides fragilis*, *Escherichia coli*, and *Enterococcus faecalis* are most often mentioned as potential participants of the process. The inflammatory process in the epithelium or deeper tissues of the intestinal wall leads to increased local blood supply. At the same time, a favorable substrate is created for the massive multiplication of bacteria, the formation of microbial biofilms, which contributes to the activation of the enzymatic systems of bacteria, increasing concentrations of potentially dangerous mutagenic products of microbial metabolism. According to [30], the highest specificity of microorganisms contributing to the occurrence of colorectal cancer is noted in streptococci such as *Streptococcus bovis* and *Streptococcus gallolyticus.* Other authors indicate a violation of homeostasis in the intestine and emphasize the role of *Lactobacillus* deficiency in reducing the protective mechanisms [28].

The analysis of statistical data shows that there is an activation of the biosynthesis of fatty acids against the background of inhibition of the biosynthesis of amino acids and glycan in patients with colorectal cancer compared with the control group [26]. Statistically significant differences in the levels of metabolites of microbial origin, namely, an increase in the relative concentrations of phenylacetic, isobutyric, valeric, isovaleric acids, and hexose-phosphates with a simultaneous decrease in taurine, glutamine, β-alanine, isoleucine, galactose, xylose, glycerol, methanol, ornithine, guanidine, choline acid, and its derivatives, 4-aminohippuric acid, have been identified in a recent paper [32].

Certainty is not currently attainable regarding the use of volatile fatty acids as markers of oncology. Reducing the levels of SCFA (acetic, butyric), secondary bile acids, concomitant increase in amino acids (leucine, valine, proline, serine) valeric, isobutyric, isovaleric acid can be associated with the activity of enzymatic systems of *Ruminococcus* spp., *Fusobacterium*, *Porphyromonas*, *Clostridia*, *Lachnospiraceae*. Changes in the composition of the microbiota in patients with colorectal cancer, noted by the authors of the review [31], can be used as a diagnostic method. Also, the review authors [33] propose to use the following compound profile: short-chain fatty acids (mainly butyric acid), cholium-kilot on deoxycholic acid derivatives, bacterial toxin fragilis, and trimethylamine-N-oxide for the diagnosis of colorectal cancer. Other authors [34] also suggest a bacterial metabolite butyric acid as a marker for colorectal cancer.

An alternative concept is that volatile fatty acids, for example, butyric acid, may have a protective effect, which slows down the development of large intestine malignancies. Butyrate-producing bacteria contained in the microbiota of the gastrointestinal tract, such as *Faecalibacterium prausnitzii*, *Eubacterium rectale*, or *Roseburia*, promote an increase in the content of butyric acid [35].

A treatment of large amounts of information on substances of bacterial origin potentially capable of being included in human metabolism allows us to distinguish six groups of compounds, based on the profile of which early diagnosis of colorectal cancer can be built [29]. There are short-chain fatty acids, bile acids, indoles, cresols, phenolic (phenyl-containing fatty) acids, and polyamines. Analysis of literature data [27] shows that under the influence of microbiota, changes in the directions of chemical transformation of glucose, fats, and amino acids are possible.

The metabolic profile, largely formed by the microbiota, was used as a diagnostic method for cancer not directly related to the gastrointestinal tract. Statistically significant differences in the content of substances involved in the metabolism of glycerol lipids and retinol and ways of ethylbenzene degradation can be used to diagnose bladder cancer. Such metabolites are actively produced and/or absorbed with the participation of enzymes of *Herbaspirillum*, *Gemella*, *Bacteroides*, *Porphyrobacter*, *Faecalibacterium*, *Aeromonas*, and *Marmoricola* [36].

A change in the metabolic profile of amino acids such as valine, cysteine, tyrosine, and 6-hydroxynicotinic acid can be used as a method for diagnosing oral cancer [37]. Substances of microbial origin and components of the metabolism of *Helicobacter pylori* have a significant impact on the formation and growth of malignant neoplasms of the esophagus, large intestine, pancreas, and lung. A cross-sectional statistical analysis shows that the likelihood of oncological complications associated with *Helicobacter pylori* increases in smokers and patients diagnosed with chronic pancreatitis and diabetes [38].

The metabolites produced by the microbiota of the upper respiratory tract and lungs may influence the development of oncological processes in them. Three types of bacteria, *Granulicatella*, *Streptococcus*, and *Veillonella*, are mentioned most often in this connection. They probably have differences in the metabolism of polyamines, expressed in elevated levels of putrescine and similar products. According to other data, dysbiosis and an increase in *Streptococcus* and *Mycobacterium* are practically not associated with the development of lung cancer [39].

However, waste products of bacteria can contribute to the development of breast cancer [22, 40–42]. The waste products of bacteria of the gastrointestinal tract can contribute to the development of malignant tumors of any other location: lung cancer [38, 39], bladder [36], pancreas [38], including hormone-dependent forms of breast cancer [22, 38, 40–42], and prostate cancer [43].

Statistically significant correlations between the levels of secondary bile acids and the incidence of breast cancer were found in [22]. The authors believe that lithocholic acid, which is a product of the metabolism of microorganisms, is able to limit the proliferation of breast cancer cells both in vitro and in vivo by activating the TGR5 receptor. Changes in the metabolism of hormones, cysteine, and methionine and the biosynthesis of fatty acids associated with breast cancer were noted in a similar study [41], but there is no definite connection between them. The search for low-molecular markers of breast cancer, carried out in [42], allowed identification of 12 compounds (amino acids, organic acids, and nucleosides) that pretend to this role. These compounds are included in the metabolism of amino acid and nucleoside metabolism.

Microbiota metabolites are able to act as accelerants and inhibitors of oncological processes. Now a scientific search in this field of knowledge is in the stage of intensive development and accumulation of a critical amount of information. The use of metabolomic approaches in combination with modern methods of statistical processing of large amounts of data undoubtedly contributes to the development of fundamental and applied medicine in the field of diagnosis and treatment of oncological diseases.

3.3 Neurological pathology and mental disorders

Some substances that form the amino acid metabolism can overcome the hemato-encephalic barrier and have a direct effect on the brain (**Figure** 7) [14, 44–46]. A search for such low-molecular compounds, quantitative determination, and their ratios can serve as the basis for the development of methods for early diagnosis, including cognitive and mental disorders [47]. It is important to note that metabolites can directly enter the region of the medulla, with blood through arteria vertebralis-arteria spinalis, bypassing the hemato-encephalic barrier, and that critical vital centers of respiration and circulation are located there.

It is unlikely that metabolites of microbial biotransformation of amino acids are the direct cause of mental or neurological diseases. At the same time, numerous experimental studies indicate the existence of a direct "intestine-microbiota-brain" link. Current evidence suggests that multiple mechanisms, including endocrine and neurocrine pathways, may be involved in gut microbiota-to-brain signaling and that the brain can in turn alter microbial composition and behavior via the autonomic nervous system [48].

The authors in literature sources traditionally attend to the aromatic amino acid tryptophan metabolism mainly due to its relationship with the synthesis of serotonin (5-HT) and melatonin [49]. Tryptophan biotransformation in humans can occur in different ways: either with the participation of endogenous enzymes that are synthesized by the intestinal cell wall or with the participation of bacterial enzymes. Accordingly, the ratios of end products of tryptophan metabolism will differ. This is easily seen by comparing the enzymes and metabolic products of tryptophan in **Figures** 5 and 8.

The traditional view is that the amino acid tryptophan is used primarily for protein synthesis or the formation of serotonin and melatonin. However, more than 90% of tryptophan was found to be metabolized into N-formyl-kynurenine followed by kynurenine (**Figure** 8) [50]. The presence of anthranilic and 3-hydroxy-anthranilic acids attracts particular attention as tryptophan metabolites. This pathway is not presented in mammalian metabolism. Such reactions of indole compounds are possible only with the participation of microbiota enzymatic systems. This also applies to picolinic and quinolinic acids, the formation of which

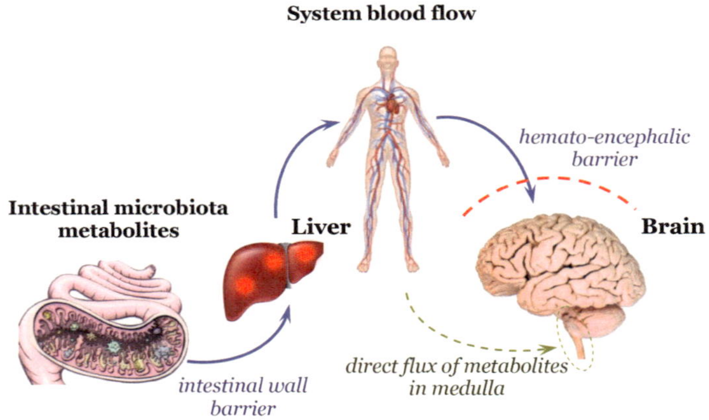

Figure 7.
Scheme of amino acid metabolite transport in the brain.

is associated with the opening of the indole ring, which can occur exclusively in the process of microbial biotransformation.

A decrease of tryptophan, xanthurenic, 3-hydroxyanthranilic, and quinolinic acids in the blood was recorded in the case of clinical occurrences of Alzheimer's disease. The same metabolites are given in [51] as potential markers of Alzheimer's disease. It can be assumed that one of the Alzheimer's disease triggers is a chronic deficiency of these substances.

Now there are two alternative hypotheses in the literature regarding products of tryptophan metabolism and their influence on the development of schizophrenia. One of them postulates that a chronic tryptophan deficiency results in failure of catabolism products, such as 3-hydroxykynurenine, quinolinic, picolinic, xanthurenic, kinureric, and anthranilic acids. Some authors maintain that such deficiency stipulates the psychosomatic symptoms of schizophrenia [52]. Other authors come to the opposite conclusion based on the analysis of statistical data [53]. They indicate a direct correlation of clinical manifestations of schizophrenia with an increased content of kynurenic acid in the cerebrospinal fluid relative to the control group. Such conflicting data emphasize once again the peculiarities of the metabolic approach. You should not limit yourself to searching and measuring one or two metabolites during clinical trials; it is important to evaluate the complex metabolic profile, to compare the indicators with positive and negative dynamics. In addition, other mechanisms that are not related to the metabolism of neurotransmitters may be the basis of mental and neurologic disorders.

Thus, attempts to search for low-molecular markers of autism [54] and depressive disorder [55] were unsuccessful. But the data indicating the potential role of the metabolism of aromatic amino acids were discovered in such a mental disorder as anorexia nervosa. Levels of tryptophan and phenylalanine were significantly reduced in patients compared with the healthy ones.

Changes in the distribution of the tryptophan metabolism products, such as kynurenine, 3-hydroxy kynurenine, kynurenic, and anthranilic acids, are observed

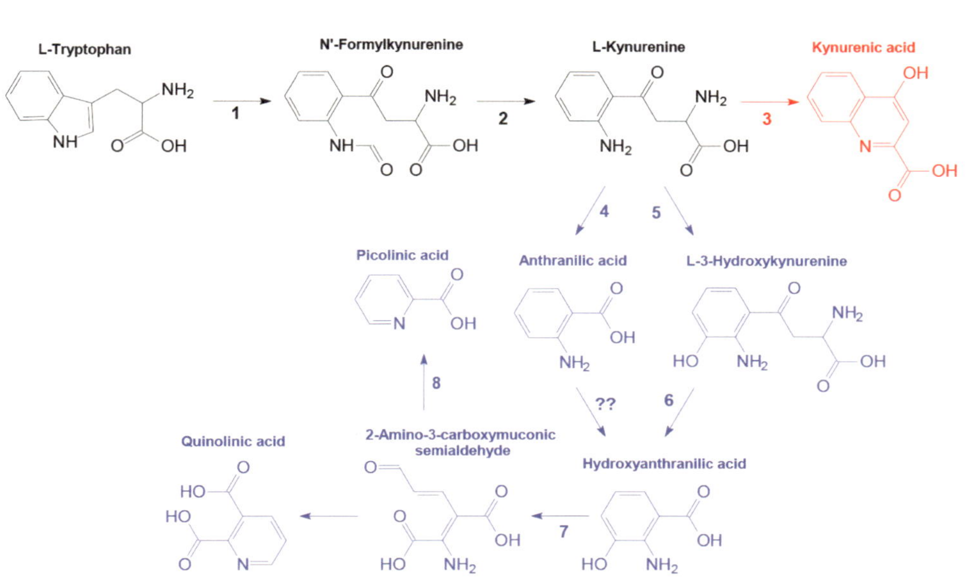

Figure 8.
Tryptophan metabolism. Enzymes: (1) indoleamine deoxygenase or tryptophan deoxygenase; (2) formidase; (3) kynurenine aminotransferase; (4) kynureninase; (5) kynurenine-3-monooxygenase; (6) kynureninase; (7) 3- hydroxyanthranilate 3,4-dioxygenase; (8) 2-amino-3- carboxymuconate-semialdehyde decarboxylase (according to [50]).

in patients with symptoms of Parkinson's disease [56]. Low levels of norepinephrine, dopamine, homovanillic acid, serotonin, and 5-hydroxyindoleacetic acid in the blood are fixed in these patients relative to the control group [57].

The failure of aromatic L-amino acid decarboxylase in combination with reduced levels of important metabolites such as serotonin, dopamine, and catecholamines leads to disruptions in the normal functioning of the whole organism, including brain activity. Crisis of oculomotor function along with muscular hypotonia and dystonia is observed in combination with other neurological syndromes in a similar state [58]. A decrease in the blood concentrations of homovanillic, 5-hydroxyindoleacetic acids, and 3-*o*-methyldopamine—substances included in the metabolism of tyrosine (**Figure 4**)—was observed in all patients. It can be noted with a high assurance that the deficiency of these metabolites is due to the lack of the transformation enzymes responsible for these reactions of the aromatic amino acids usually found in the microbiota.

3.4 Prospects for neurorehabilitation

Scientists have used metabolomics to gain new knowledge about the significance of the role that bacteria play in complex regulatory processes of higher nervous activity. Understanding the potential for managing this process cannot leave psychiatrists, neurologists, and neurorehabilitation specialists indifferent [59–61]. This fact is due to the relevance and high frequency of pathology of the nervous system. Prospects for the correction of microbiota metabolism for neurorehabilitation and the demand for this scientific search for new solutions in this area cannot be overestimated.

One of the areas discussed in the literature is the transformation of the species composition of the patient's microbiota to eliminate the deficiency of certain microorganisms. This idea has a scientific ground that many bacteria from the human microbiota in the in vitro study revealed the ability to produce hormones and neurotransmitters, that is, the presence of appropriate enzyme systems. These data are summarized in the reviews [44, 62] and in brief form are presented in **Table 2**.

Certain reports indicate that the treatment with large doses of *Lactobacillus casei* has a positive effect. Patients with chronic fatigue syndrome reported a decreased strain ($n = 39$). Patients who took a probiotic reported a significant decrease in symptoms of anxiety and had a substantial increase in the number of *Lactobacillus* and *Bifidobacteria* compared with the control group ($p = 0.01$). [63]. At the same time, treatment with live microorganisms (including fecal microbiota transplantation, FMT) is hardly predictable and can have negative consequences due to the variability of bacterial metabolism depending on the environment. For example, a randomized, double-blind, controlled study on the use of a drug based on lactobacilli in combination with prebiotic gives a negative result in patients with pancreatic necrosis: the mortality rate in the group receiving the biological product was significantly higher than in the control [64].

Neurorehabilitation of patients in modern clinics is considered as a component of acute cerebral therapy and starts from the earliest periods after injuries, strokes, and brain operations, even at the stage of the patient's stay in the intensive care unit. This is a multicomponent and long-term process aimed not only at saving lives but also at restoring motor activity, correcting neuro-endocrine, cognitive impairments, and emotional status. Different methods of monitoring the effectiveness of intensive care and the rehabilitation of the functional state of patients with various brain injuries are used [65].

The authors of this chapter believe that neurorehabilitation can be significantly enriched with a set of targeted measures aimed at correcting disorders in the development of which metabolic products associated with microbiota are actively involved. Our accumulated data on the magnitude of changes in the profile of

Hormone, neurotransmitter	Bacteria
Norepinephrine	*Bacillus subtilis*, *Bacillus mycoides*, *Proteus vulgaris*, *Serratia marcescens*
Dopamine	*B. subtilis*, *B. mycoides*, *Bacillus cereus*, *Staphylococcus aureus*, *P. vulgaris*, *S. marcescens*, *Escherichia coli*, *Morganella morganii*, *Klebsiella pneumonia*, *Hafnia alvei*, *Lactobacillus helveticus*, *Lactobacillus delbrueckii* subsp. bulgaricus
Dopamine precursor (DOPA)	*E. coli*, *B. cereus*, *L. helveticus*, *L. casei*, *L. delbrueckii* subsp. bulgaricus, *Toxoplasma gondii*
Serotonin	*S. aureus*, *Enterococcus faecalis*, *Rhodospirillum rubrum*, *B. subtilis*, *E. coli*, *M. morganii*, *K. pneumonia*, *H. alvei*, *Lactococcus lactis* subsp. cremoris, *L. lactis* subsp. lactis, *Lactococcus plantarum*, *L. helveticus*
Histamine	*M. morganii*, *P. vulgaris*, *Proteus mirabilis*, *Klebsiella* sp., *Enterobacter aerogenes*, *E. cloacae*, *Citrobacter freundii*, *Enterobacter amnigenus*, *Vibrio alginolyticus*, *Acinetobacter lowfli*, *Pseudomonas fluorescens*, *P. putida*, *Aeromonas* spp., *Clostridium* spp., *Photobacterium* spp., *Lactobacillus buchneri*, *Streptococcus thermophilus*
γ-Aminobutyric acid	*Bifidobacterium adolescentis*, *B. dentium*, *B. infantis*, *B. angulatum*, *Lactobacillus brevis*, *L. plantarum*, *L. paracasei*, *L. buchneri*, *L. helveticus*, *L. delbrueckii*, *L. reuteri*, *L. zymae*
Tyramine	*Lactobacillus* spp., *Lactococcus* spp., *Enterococcus* spp., *Carnobacterium*

Table 2.
Literary data on the ability of many bacteria: representatives of the human microbiota to participate in the production of hormones and neurotransmitters (adapted from [44, 62]).

microbiota metabolites and their connection with the course and outcome of the disease in patients with lesions of the central nervous system indicate the possibility of their use in choosing tactics for managing patients with this pathology. This complex may include several areas: (i) the first is the additional introduction into the body of substances that are associated with a shortage of other clinical manifestations of pathology. This can be achieved by nutritional correction or dietary supplements, including those obtained using industrial microbiology methods, as well as the administration of parenteral preparations containing the necessary metabolites of microbial origin. (ii) The second is the suppression of the metabolic activity of those types of bacteria in the composition of the microbiota, which in excess produce "unwanted" metabolites, through the selective use of antibacterial drugs with an appropriate mechanism of action. (iii) The third is the elimination of excess unwanted metabolites in the systemic circulation through the targeted use of extracorporeal blood purification procedures with filters/sorbents that remove specific substances.

Of course, the use of modern metabolic methods for an objective assessment of the dynamics of the profile of metabolites in parallel with the monitoring of the psychosomatic state, functions of the damaged brain, spasticity level, motor skills, etc. is necessary for the successful implementation of the above directions in a particular patient. But above all, reliable data on key microbial metabolites, the level of which must be monitored in patients in the process of neurorehabilitation to on must be obtained. For example, metabolites associated with the development of septic shock (p-HPhAA) [66, 67] and death (PhA, p-HPhLA) [10] were earlier established for patients with sepsis. At the same time, another metabolite—PhPA—was a characteristic for the metabolic profile of a healthy person. The study of metabolome is conducted using the GC–MS method for patients with affection of the central nervous system of various etiologies [68]. Currently, the purpose of this study is to detect microbial metabolites associated with changes in the neurological status of patients in the process of neurorehabilitation. Preliminary results indicate a number of significant features, for example, positive neurological and psychosomatic dynamics is associated with the appearance and accumulation of the metabolite p-HBA in the intestine and the patient's blood, which is not observed in other groups of patients. The composition of the microbiota in patients with severe neurosomatic pathology using the method of metagenomic sequencing of the 16S pRNA is under study. Correlations with microbial blood metabolites are also being studied. Preliminary data demonstrate significant differences when comparing various patient groups [69]. The results of the multicenter study will serve as the basis for the development and objective evaluation of the effectiveness of the above technologies in the process of neurorehabilitation.

4. Conclusion

A new level of knowledge about the role of the microbiota in the human body was made possible by metabolomics. In the coming years, this will lead to new solutions in the diagnosis of many "difficult" diseases. Methods of active control of metabolic processes that will subordinate the dysfunction of the "invisible organ" to the benefit of the host will be found. It will lead to the increase in the effectiveness of treatment and successful rehabilitation of patients. In particular, in the field of neurorehabilitation, clinical studies are currently aimed at finding such methods for correcting the metabolism of microbiota that will achieve a balance of low-molecular metabolites as signaling molecules of microbiota to restore brain function.

Acknowledgements

This work was supported by the Russian Science Foundation Grant № 15–15-00110-P.

Author details

Natalia V. Beloborodova[1*], Andrey V. Grechko[1] and Andrey Yu Olenin[2]

1 Federal Research and Clinical Center of Intensive Care Medicine and Rehabilitology, Moscow, Russia

2 Department of Chemistry, M.V. Lomonosov Moscow State University, Moscow, Russia

*Address all correspondence to: nvbeloborodova@yandex.ru

References

[1] Chernevskaya E, Beloborodova N. Gut microbiome in critical illness (review). General Reanimatology. 2018;**14**:96. DOI: 10.15360/1813-9779-2018-5-96-119

[2] Wishart DS, Feunang YD, Marcu A, Guo AC, Liang K, Vázquez-Fresno R, et al. HMDB 4.0: The human metabolome database for 2018. Nucleic Acids Research. 2018;**46**:D608. DOI: 10.1093/nar/gkx1089

[3] Hornung B, dos Santos VAPM, Smidt H, Schaap PJ. Studying microbial functionality within the gut ecosystem by systems biology. Genes and Nutrition. 2018;**13**:5. DOI: 10.1186/s12263-018-0594-6

[4] Hooper LV, Macpherson AJ. Immune adaptations that maintain homeostasis with the intestinal microbiota. Nature Reviews. Immunology. 2010;**10**:159. DOI: 10.1038/nri2710

[5] Cantarel BL, Lombard V, Henrissat B. Complex carbohydrate utilization by the healthy human microbiome. PLoS One. 2012;**7**:e28742. DOI: 10.1371/journal.pone.0028742

[6] Kim CH. Immune regulation by microbiome metabolites. Immunology. 2018;**154**:220. DOI: 10.1111/imm.12930

[7] HMP Consortium. Structure, function and diversity of the healthy human microbiome. Nature. 2012;**486**:207. DOI: 10.1038/nature11234

[8] Gill SR, Pop M, DeBoy RT, Eckburg PB, Turnbaugh PJ, Samuel BS, et al. Metagenomic analysis of the human distal gut microbiome. Science. 2006;**312**:1355. DOI: 10.1126/science.1124234

[9] Yadav M, Verma MK, Chauhan NS. A review of metabolic potential of human gut microbiome in human nutrition. Archives of Microbiology. 2018;**200**:203. DOI: 10.1007/s00203-017-1459-x

[10] Beloborodova NV. Interaction of host-microbial metabolism in sepsis. In: Kumar V, editor. Sepsis. InTechOpen; 2017. pp. 3-19. DOI: 10.5772/68046

[11] Gertsman I, Gangoiti JA, Nyhan WL, Barshop BA. Perturbations of tyrosine metabolism promote the indolepyruvate pathway via tryptophan in host and microbiome. Molecular Genetics and Metabolism. Elsevier. 2015;**114**:431. DOI: 10.1016/j.ymgme.2015.01.005

[12] Land EJ, Ramsden CA, Riley PA. Pulse radiolysis studies of ortho-quinone chemistry relevant to melanogenesis. Journal of Photochemistry and Photobiology. B. 2001;**64**:123. DOI: 10.1016/S1011-1344(01)00220-2

[13] Faccio G, Kruus K, Saloheimo M, Thöny-Meyer L. Bacterial tyrosinases and their applications. Process Biochemistry. 2012;**47**:1749. DOI: 10.1016/j.procbio.2012.08.018

[14] van den Brink WJ, Palic S, Köhler I, de Lange ECM. Access to the CNS: Biomarker strategies for dopaminergic treatments. Pharmaceutical Research. 2018;**35**:64. DOI: 10.1007/s11095-017-2333-x

[15] Wilson ID, Nicholson JK. Gut microbiome interactions with drug metabolism, efficacy, and toxicity. Translational Research. 2017;**179**:204. DOI: 10.1016/j.trsl.2016.08.002

[16] Nyangale EP, Mottram DS, Gibson GR. Gut microbial activity, implications for health and disease: The potential role of metabolite analysis. Journal of Proteome Research. 2012;**11**:5573. DOI: 10.1021/pr300637d

[17] Bussche JV, Marzorati M, Laukens D, Vanhaecke L. Validated high resolution mass spectrometry-based approach for metabolomic fingerprinting of the human gut phenotype. Analytical Chemistry. 2015;**87**:10927. DOI: 10.1021/acs.analchem.5b02688

[18] Ilinskaya ON, Ulyanova VV, Yarullina DR, Gataullin IG. Secretome of intestinal bacilli: A natural guard against pathologies. Frontiers in Microbiology. 2017;**8**:1666. DOI: 10.3389/fmicb.2017.01666

[19] Han H, Li Y, Fang J, Liu G, Yin J, Li T, et al. Gut microbiota and type 1 diabetes. International Journal of Molecular Sciences. 2018;**19**:995. DOI: 10.3390/ijms19040995

[20] Zhao L, Zhang F, Ding X, Wu G, Lam YY, Wang X, et al. Gut bacteria selectively promoted by dietary fibers alleviate type 2 diabetes. Science. 2018;**359**:1151. DOI: 10.1126/science.aao5774

[21] Sitkin SI, Vakhitov TY, Demyanova EV. Microbiome, gut dysbiosis and inflammatory bowel disease: That moment when the function is more important than taxonomy. Almanac of Clinical Medicine. 2018;**46**:396. DOI: 10.18786/2072-0505-2018-46-5-396-425

[22] Mikó E, Vida A, Kovács T, Ujlaki G, Trencsényi G, Márton J, et al. Lithocholic acid, a bacterial metabolite reduces breast cancer cell proliferation and aggressiveness. BBA Bioenergitics. 2018;**1859**:958. DOI: 10.1016/j.bbabio.2018.04.002

[23] Mukherjee S, Joardar N, Sengupta S, Babu SPS. Gut microbes as future therapeutics in treating inflammatory and infectious diseases: Lessons from recent findings. The Journal of Nutritional Biochemistry. 2018;**61**:111. DOI: 10.1016/j.jnutbio.2018.07.010

[24] Alexander JL, Scott AJ, Pouncey AL, Marchesi J, Kinross J, Teare J. Colorectal carcinogenesis: An archetype of gut microbiota-host interaction. Ecancermedicalscience. 2018;**12**:865. DOI: 10.3332/ecancer.2018.865

[25] Roy S, Trinchieri G. Microbiota: A key orchestrator of cancer therapy. Nature Reviews. Cancer. 2017;**17**:271. DOI: 10.1038/nrc.2017.13

[26] Allali I, Boukhatem N, Bouguenouch L, Hardi H, Boudouaya HA, Cadenas MB, et al. Gut microbiome of Moroccan colorectal cancer patients. Medical Microbiology and Immunology. 2018;**207**:211. DOI: 10.1007/s00430-018-0542-5

[27] Zhou CB, Fang JY. The regulation of host cellular and gut microbial metabolism in the development and prevention of colorectal cancer. Critical Reviews in Microbiology. 2018;**44**:436. DOI: 10.1080/1040841X.2018.1425671

[28] Coleman OI, Haller D. Bacterial signaling at the intestinal epithelial interface in inflammation and cancer. Frontiers in Immunology. 2018;**8**:1927. DOI: 10.3389/fimmu.2017.01927

[29] Wang QQ, Li L, Xu R. A systems biology approach to predict and characterize human gut microbial metabolites in colorectal cancer. Scientific Reports. 2018;**8**:6225. DOI: 10.1038/s41598-018-24315-0

[30] Han S, Gao J, Zhou Q, Liu S, Wen C, Yang X. Role of intestinal flora in colorectal cancer from the metabolite perspective: A systematic review. Cancer Management and Research. 2018;**10**:199. DOI: 10.2147/CMAR.S153482

[31] Villéger R, Lopès A, Veziant J, Gagnière J, Barnich N, Billard E, et al. Microbial markers in colorectal cancer detection and/or prognosis. World Journal of Gastroenterology.

2018;**24**:2327. DOI: 10.3748/wjg.v24.i22.2327

[32] Gall GL, Guttula K, Kellingray L, Tett AJ, ten Hoopen R, Kemsley KE, et al. Metabolite quantification of fecal extracts from colorectal cancer patients and healthy controls. Oncotarget. 2018;**9**:33278

[33] Zou S, Fang L, Lee MH. Dysbiosis of gut microbiota in promoting the development of colorectal cancer. Gastroenterology Report. 2018;**6**:1. DOI: 10.1093/gastro/gox031

[34] Wu X, Wu Y, He L, Wu L, Wang X, Liu Z. Effects of the intestinal microbial metabolite butyrate on the development of colorectal cancer. Journal of Cancer. 2018;**9**:2510. DOI: 10.7150/jca.25324

[35] McNabney SM, Henagan TM. Short chain fatty acids in the colon and peripheral tissues: A focus on butyrate, colon cancer, obesity and insulin resistance. Nutrients. 2017;**9**:1348. DOI: 10.3390/nu9121348

[36] Wu P, Zhang G, Zhao J, Chen J, Yang C, Huang W, et al. Profiling the urinary microbiota in male patients with bladder cancer in China. Frontiers in Cellular and Infection Microbiology. 2018;**8**(167)

[37] Xie GX, Chen TL, Qiu YP, Shi P, Zheng XJ, Su MM, et al. Urine metabolite profiling offers potential early diagnosis of oral cancer. Metabolomics. 2012;**8**:220. DOI: 10.1007/s11306-011-0302-7

[38] Meng C, Bai C, Brown TD, Hood LE, Tian Q. Human gut microbiota and gastrointestinal cancer. Genomics, Proteomics and Bioinformatics. 2018;**16**:33. DOI: 10.1016/j.gpb.2017.06.002

[39] Mur LAJ, Huws SA, Cameron SJS, Lewis PD, Lewis KE. Lung cancer: A new frontier for microbiome research and clinical translation. Ecancermedicalscience. 2018;**12**:866. DOI: 10.3332/ecancer.2018.866

[40] Zhang A-h, Sun H, Qiu S, Wang XJ. Metabolomics in noninvasive breast cancer. Clinica Chimica Acta. 2013;**424**:3. DOI: 10.1016/j.cca.2013.05.003

[41] Fernández MF, Reina-Pérez I, Astorga JM, Rodríguez-Carrillo A, Plaza-Díaz J, Fontana L. Breast cancer and its relationship with the microbiota. International Journal of Environmental Research and Public Health. 2018;**15**:1747. DOI: 10.3390/ijerph15081747

[42] Chen Y, Zhang R, Song Y, He J, Sun J, Bai J, et al. RRLC-MS/MS-based metabonomics combined with in-depth analysis of metabolic correlation network: Finding potential biomarkers for breast cancer. The Analyst. 2009;**134**:2003. DOI: 10.1039/b907243h

[43] Porter CM, Shrestha E, Peiffer LB, Sfanos KS. The microbiome in prostate inflammation and prostate cancer. Prostate Cancer and Prostatic Diseases. 2018;**21**:345. DOI: 10.1038/s41391-018-0041-1

[44] Averina OV, Danilenko VN. Human intestinal microbiota: Role in development and functioning of the nervous system. Microbiology. 2017;**86**:1-18. DOI: 10.1134/S0026261717010040

[45] Umbrello G, Esposito S. Microbiota and neurologic diseases: Potential effects of probiotics. Journal of Translational Medicine. 2016;**14**:298. DOI: 10.1186/s12967-016-1058-7

[46] Willemsen MA, Verbeek MM, Kamsteeg E-J, de Rijk-van Andel JF, Aeby A, Blau N, et al. Tyrosine hydroxylase deficiency: A treatable disorder of brain catecholamine

biosynthesis. Brain. 2010;**133**:1810. DOI: 10.1093/brain/awq087

[47] Sadok I, Gamian A, Staniszewska MM. Chromatographic analysis of tryptophan metabolites. Journal of Separation Science. 2017;**40**:3020. DOI: 10.1002/jssc.201700184

[48] Mayer EA, Tillisch K, Gupta A. Gut/brain axis and the microbiota. The Journal of Clinical Investigation. 2015;**125**:926. DOI: 10.1172/JCI76304

[49] Kałużna-Czaplińska J, Gątarek P, Chirumbolo S, Chartrand MS, Bjørklund G. How important is tryptophan in human health? Critical Reviews in Food Science and Nutrition. 2019;**59**:72. DOI: 10.1080/10408398.2017

[50] Chatterjee P, Goozee K, Lim CK, James I, Shen K, Jacobs KR, et al. Alterations in serum kynurenine pathway metabolites in individuals with high neocortical amyloid-β load: A pilot study. Scientific Reports. 2018;**8**:8008. DOI: 10.1038/s41598-018-25968-7

[51] Lv C, Li Q, Liu X, He B, Sui Z, Xu H, et al. Determination of catecholamines and their metabolites in rat urine by ultra-performance liquid chromatography–tandem mass spectrometry for the study of identifying potential markers for Alzheimer's disease. Journal of Mass Spectrometry. 2015;**50**:354. DOI: 10.1002/jms.3536

[52] Kanchanatawan B, Sirivichayakul S, Thika S, Ruxrungtham K, Carvalho AF, Geffard M, et al. Physio-somatic symptoms in schizophrenia: Association with depression, anxiety, neurocognitive deficits and the tryptophan catabolite pathway. Metabolic Brain Disease. 2017;**32**:1003. DOI: 10.1007/s11011-017-9982-7

[53] Erhardt S, Schwieler L, Nilsson L, Linderholm K, Engberg G. The kynurenic acid hypothesis of schizophrenia. Physiology and Behavior. 2007;**92**:203. DOI: 10.1016/j.physbeh.2007.05.025

[54] Kałużna-Czaplińska J, Żurawicz E, Jóźwik J. Chromatographic techniques coupled with mass spectrometry for the determination of organic acids in the study of autism. Journal of Chromatography B. 2014;**964**:128. DOI: 10.1016/j.jchromb.2013.10.026

[55] Zheng P, Wang Y, Chen L, Yang D, Meng H, Zhou D, et al. Identification and validation of urinary metabolite biomarkers for major depressive disorder. Molecular and Cellular Proteomics. 2013;**12**:207. DOI: 10.1074/mcp.M112.021816

[56] Havelund JF, Andersen AD, Binzer M, Blaabjerg M, Heegaard NHH, Stenager E, et al. Changes in kynurenine pathway metabolism in Parkinson patients with L-DOPA-induced dyskinesia. Journal of Neurochemistry. 2017;**142**:756. DOI: 10.1111/jnc.14104

[57] Sitte HH, Pifl C, Rajput AH, Hörtnagl H, Tong J, Lloyd GK, et al. Dopamine and noradrenaline, but not serotonin, in the human claustrum are greatly reduced in patients with Parkinson's disease: Possible functional implications. The European Journal of Neuroscience. 2017;**45**:192. DOI: 10.1111/ejn.13435

[58] Manegold C, Hoffmann GF, Degen I, Ikonomidou H, Knust A, Laaß MW, et al. Aromatic L-amino acid decarboxylase deficiency: Clinical features, drug therapy and follow-up. Journal of Inherited Metabolic Disease. 2009;**32**:371. DOI: 10.1007/s10545-009-1076-1

[59] Dovrolis N, Kolios G, Spyrou GM, Maroulakou I. Computational profiling of the gut-brain axis: *Microflora dysbiosis* insights to neurological disorders. Briefings in Bioinformatics. 2017. bbx154. DOI: 10.1093/bib/bbx154

[60] Carabotti M, Scirocco A, Maselli MA, Severi C. The gut-brain axis: Interactions between enteric microbiota, central and enteric nervous systems. Annals of Gastroenterology. 2015;**28**:203

[61] Singh V, Roth S, Llovera G, Sadler R, Garzetti D, Stecher B, et al. Microbiota dysbiosis controls the neuroinflammatory response after stroke. The Journal of Neuroscience. 2016;**36**:7428. DOI: 10.1523/jneurosci.1114-16.2016

[62] Lucas P, Landete J, Coton M, Coton E, Lonvaud-Funel A. The tyrosine decarboxylase operon of *Lactobacillus brevis* IOEB 9809: Characterization and conservation in tyramine-producing bacteria. FEMS Microbiology Letters. 2003;**229**:65. DOI: 10.1016/S0378-1097(03)00787-0

[63] Rao V, Bested AC, Beaulne TM, Katzman MA, Iorio C, Berardi JM, et al. A randomized, double-blind, placebo-controlled pilot study of a probiotic in emotional symptoms of chronic fatigue syndrome. Gut Pathogens. 2009;**1**:6. DOI: 10.1186/1757-4749-1-6

[64] Besselink MG, van Santvoort HC, Buskens E, Boermeester MA, van Goor H, Timmerman HM, et al. Probiotic prophylaxis in predicted severe acute pancreatitis: A randomized, double-blind, placebo-controlled trial. Lancet. 2008;**371**:651. DOI: 10.1016/s0140-6736(08)60207-x

[65] Kiryachkov YY, Grechko AV, Kolesov DL, Loginov AA, Petrova MV, Rubanes M, et al. Monitoring of the effectiveness of intensive care and rehabilitation by evaluating the functional activity of the autonomic nervous system in patients with brain damage. Obshchaya Reanimatologiya. 2018;**14**(4):21. DOI: 10.15360/1813-9779-2018-4-21-34

[66] Beloborodova NV, Olenin AY, Pautova AK. Metabolomic findings in sepsis as a damage of host-microbial metabolism integration. Journal of Critical Care. 2018;**43**:246. DOI: 10.1016/j.jcrc.2017.09.014

[67] Beloborodova NV, Sarshor YN, Bedova AY, Chernevskaya EA, Pautova AK. Involvement of aromatic metabolites in the pathogenesis of septic shock. Shock. 2018;**50**:273. DOI: 10.1097/shk.0000000000001064

[68] Pautova AK, Bedova AY, Sarshor YN, Beloborodova NV. Determination of aromatic microbial metabolites in blood serum by gas chromatography–mass spectrometry. Journal of Analytical Chemistry. 2018;**73**:160. DOI: 10.1134/S1061934818020089

[69] Beloborodova NV, Chernevskaya EA, Pautova AK, Bedova AY, Sergeev AA. Altered serum profile of aromatic metabolites reflects the biodiversity reduction of gut microbiota in critically ill patients. Critical Care. 2018;**22**(Suppl 1):82. DOI: 10.1186/s13054-018-1973-5

Chapter 4

Serum Metabolomics as a Powerful Tool in Distinguishing Trauma from Other Critical Illness Conditions

Elizabeth R. Lusczek

Abstract

Critical illness is highly variable, complicating patient care and recovery. We have previously used metabolomics to investigate several causes of intensive care unit admission, seeking to assess changes in metabolism occurring with each condition. We present a meta-analysis of these serum metabolomes, exploring how the metabolomes differ with each condition. We also present how mass spectrometry-based metabolomics could be used for predictive monitoring. Serum metabolites were previously quantified using nuclear magnetic resonance spectroscopy in patients with traumatic injury, respiratory failure, pancreatitis, and combat trauma. Healthy controls are also included. Spectral features were analyzed with principal component analysis (PCA) to explore patterns in patients' underlying conditions. PCA suggests trauma metabolic profiles, particularly combat casualties, differ from other conditions. Principal components 2 and 3, accounting for 16% of the variation in the model, distinguish samples obtained from trauma patients. Metabolomics is a powerful tool for quantifying variability in critical illness, highlighting trauma as separate from other conditions. This observation is in line with the -omics literature, which has described a massive global "genomic storm" in response to severe injury. Mass spectrometry highlights this extreme variability, which occurs in ICU patients but not healthy controls. With new technology, metabolomics could be used to bring faster, individualized patient care to the ICU.

Keywords: metabolomics, NMR, ICU, critical illness, biomarker, traumatic injury, combat casualty, mass spectrometry

1. Introduction

Critical illness encompasses a wide variety of life-threatening conditions, often requiring intensive monitoring and sophisticated life support, such as dialysis, mechanical ventilation, and nutritional support. Patients are cared for in intensive care units (ICUs), staffed by specialists. Because patients' conditions can change quickly over time, ICU staff are highly trained and nurses regularly care for only one or two patients at a time. Because of these factors, critical illness carries a high cost burden. It has been estimated that anywhere from 17 to 39% of hospital costs in

IntechOpen

the United States are due to critical illness. Total costs, including 1 year of care after discharge are estimated at $121–263 billion, or 5–11% of United States health care expenditures [1]. The cost burden is difficult to estimate, due in part to the complicated recovery process.

Recently, post-intensive care syndrome (PICS) has been identified as a constellation of cognitive, psychological, and physical impairments that result from critical illness [2], occurring with increased prevalence due to the increased survivability of critical illness [3]. ICU-acquired delirium and mechanical ventilation are among the risk factors for PICS, and the effects can be long-lasting. An estimated 90% of patients report ICU-acquired weakness lasting 2–5 years from ICU discharge, and 74% of patients with acute respiratory distress syndrome report cognitive impairments at discharge. Approximately a quarter of these patients report effects lasting as long as 6 years [4].

While survivability from critical illness has increased, it has been difficult to make further advances in patient care and outcomes due to the heterogeneity of the patient population. Respiratory disorders requiring mechanical ventilation, acute myocardial infarction, intracranial hemorrhage, percutaneous cardiovascular procedure with drug-eluting stent, and septicemia are the leading causes of ICU admission, but gastrointestinal disorders, renal disorders, and trauma are also frequent causes of ICU admission [5]. To further complicate matters, as many as 1/3 of ICU patients have multiple co-morbidities. Homogenous patient populations can be difficult to identify, let alone study, in the ICU. As such, a "one-size-fits-all" approach to patient care can lead to unpredictable results. To cope with these hallmarks of critical illness, modern ICU clinicians argue for precision medicine approaches to critical care as a way to improve patient care [6–8].

Metabolomics, which reflects the phenome more closely than other -omics disciplines, may be a key to this endeavor. This terrain has been largely unexplored, save for a few studies. Targeted metabolomics has been used to discriminate non-infectious systemic inflammatory response syndrome (SIRS) from infections SIRS [9]. Untargeted approaches have identified significant, severe metabolic derangements that are associated with mortality [10, 11].

This chapter presents efforts to use metabolomics to explore this difficult-to-study space. Namely, critical illness is highly variable and affects diffuse organ systems in a heterogeneous patient population that may have multiple co-morbidities. Since the metabolome is closest to the phenome, it is more likely to reflect the individual patient's state at any given time than other -omes. As others have pointed out, issues of heterogeneity and variability make biomarker studies problematic [6, 10]. A first step is to examine how metabolic profiles differ with different underlying diseases and with illness severity to get a better sense of this variability. This chapter touches on current efforts in this direction.

2. Metabolomics methodology and previous work

The NMR-based metabolomics studies we performed were pilot studies seeking to characterize metabolic profiles in combat injury [12], civilian traumatic injury [13], acute pancreatitis [14], and respiratory failure [15]. Healthy controls were also profiled [12, 13].

The use of the same protocol to process serum samples, collect NMR spectra, and quantify metabolites allows for a meta-analysis comparing the metabolic profiles from each study.

Briefly, samples were filtered using a 3 kDa ultracentrifuge filter to remove large molecules such as proteins that bind to the internal standard. Filtrate is mixed in equal parts with 200 mM sodium phosphate buffer and with 50 microliters of the internal standard 3-(trimethylsilyl)propionic acid. A 1D Carr-Purcell-Meiboom-Gill (CPMG) pulse sequence was used to collect spectra, and metabolites were identified and quantified using Chenomx software [16]. Full experimental details can be found in the original research articles [12–15].

Metabolic profiles were limited to the 41 metabolites identified and quantified in all four studies. Metabolite concentrations (millimoles per liter or mM) were log-transformed and auto-scaled before principal component analysis (PCA) was performed with R software [17]. PCA scores were colored by underlying diagnosis/patient group (combat trauma, civilian trauma age 21–40, civilian trauma age 65 and older, acute pancreatitis, respiratory failure, healthy controls age 21–40, and healthy controls age 65 and older).

For the purposes of visualizing the diagnosis groups in this meta-analysis, some simplifications were made based on the previously published results. Because no clear difference was seen between patients in respiratory failure regardless of underlying cause, patients with chronic obstructive pulmonary disease (COPD) exacerbation, heart failure, and pneumonia were combined into the “respiratory failure” group [15]. Non-hospitalized patients who did not develop pancreatitis [14] and non-hospitalized patients with stable COPD [15] were excluded from this analysis to facilitate visualization.

3. Meta-analysis results

In total, 291 serum samples were analyzed with principal component analysis. Most of these were from trauma patients. The number of samples studied in each diagnosis group is presented in **Table 1**.

Principal component analysis scores (**Figures 1** and **2**) and loadings (**Figures 3** and **4**) are shown for the first three components. Each dot in the scores plot represents a serum sample, which is colored according to the diagnosis or condition. The loadings plots show how the metabolites profiled contribute to the model. The first three components account for 51% of the variability in the data. Component 1 accounts for 35% of the variation; components 2 and 3 account for 9.8 and 6.6% of the variation, respectively. A three-dimensional biplot (**Figure 5**) helps visualize all the information.

Interestingly, the most meaningful pattern in the PCA scores is observed in **Figure 2**, the plot of component 2 vs. component 3. A clear line can be drawn along PC2 and PC3, demarcating the samples from trauma patients (red,

Condition	Number of samples
Combat Trauma	111
Civilian Trauma (age 21-40)	36
Civilian Trauma (age 65+)	42
Respiratory Failure	23
Acute Pancreatitis (hospitalized)	30
Healthy Control (age 21-40)	23
Healthy Control (age 65+)	26

Table 1.
Number of samples profiled with NMR-based metabolomics per condition studied.

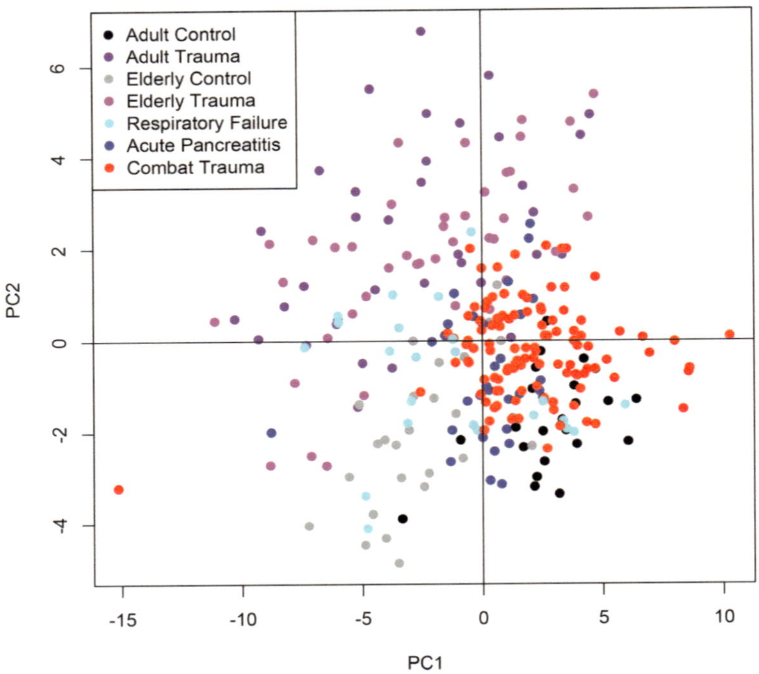

Figure 1.
Scores plot of PC1 vs. PC2 for serum samples described in ***Table 1****. Samples are colored by diagnosis.*

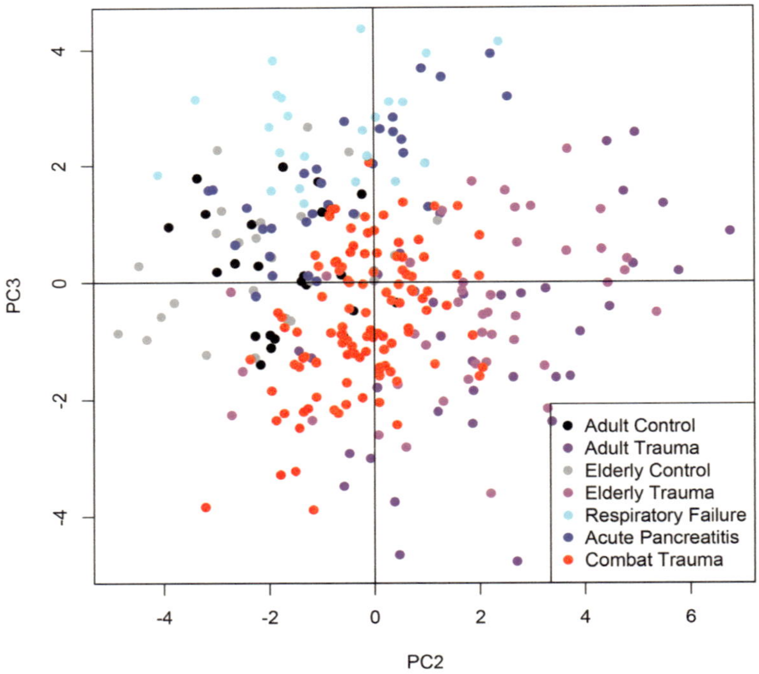

Figure 2.
Scores plot of PC2 vs. PC3 for serum samples described in ***Table 1****. Samples are colored by diagnosis. These two principal components most clearly distinguish trauma samples from non-trauma samples.*

DOI: http://dx.doi.org/10.5772/intechopen.87145

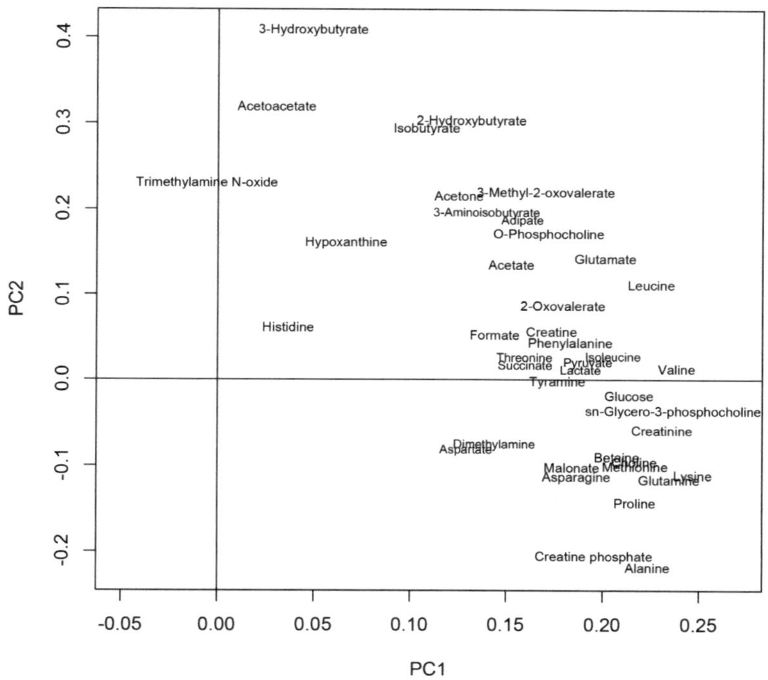

Figure 3.
*Loadings plot of PC1 vs. PC2 for serum samples described in **Table 1**. Loadings values are shown in **Table 2**.*

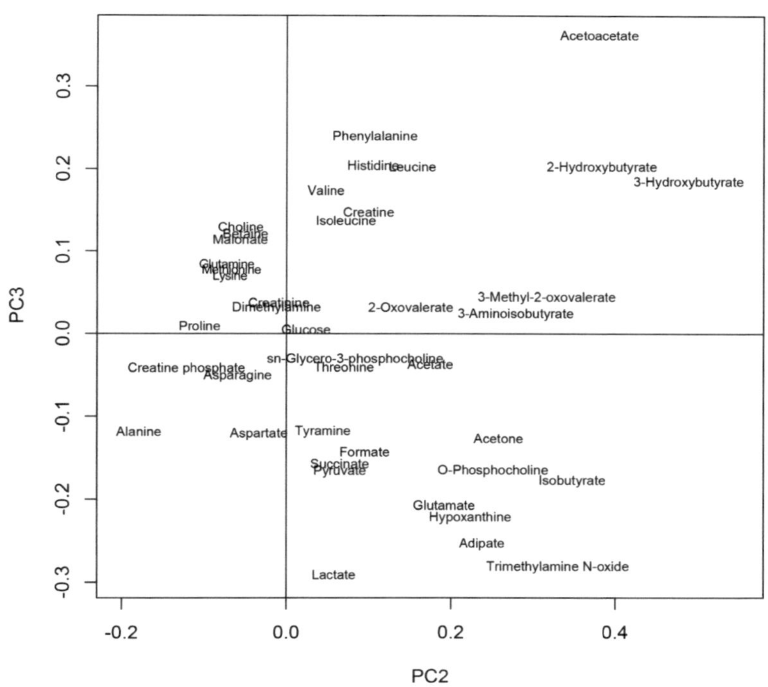

Figure 4.
*Loadings plot of PC2 vs. PC3 for serum samples described in **Table 1**. Loadings values are shown in **Table 2**, and the magnitude of the loadings vector spanned by PC2 and PC3 is calculated. Metabolites most associated with trauma occupy the lower right quadrant.*

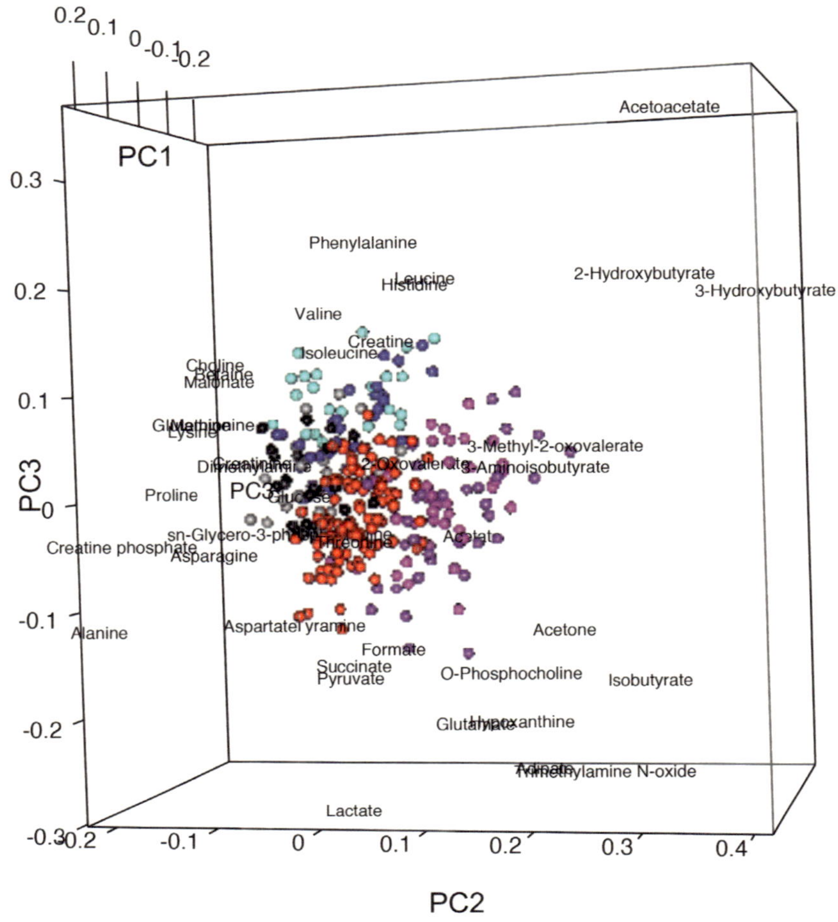

Figure 5.
*Biplot of the first three principal components for serum samples described in **Table 1**. Samples are colored by diagnosis.*

purple, or magenta) from the healthy controls (gray or black) and the patients with other conditions (blue or light blue).

The loadings vectors for the first three principal components are reported in **Table 2**. Since PC2 and PC3 can be used to discriminate trauma samples from non-trauma samples, we used the loadings of these components to identify the metabolites most associated with trauma. To do this, we calculated the magnitude of the vector spanned by PC2 and PC3, shown in column 4 of **Table 2**. The 10 metabolites with the largest magnitude in the PC loadings are acetoacetate, 3-hydroxybutyrate, trimethylamine N-oxide, 2-hydroxybutyrate, isobutyrate, adipate, lactate, hypoxanthine, glutamate, and alanine. These metabolites reflect disruptions to energy metabolism and oxidative stress.

4. Meta-analysis discussion

Our metabolomics studies can be united under a common theme: all were done in conditions that are common causes for admission to the ICU. Because sample preparation protocol is the same for all our serum-based NMR metabolomics

Metabolite	PC1 Loading	PC2 Loading	PC3 Loading	Magnitude of PC2 and PC3
Acetoacetate	0.00380019	0.31775843	0.36065228	0.48066671
3-Hydroxybutyrate	0.01446508	0.40722582	0.18306298	0.4464806
TMAO	-0.0488827	0.22963983	-0.2823471	0.36394273
2-Hydroxybutyrate	0.09672712	0.30185385	0.20091817	0.36260702
Isobutyrate	0.08468395	0.29280528	-0.1782326	0.34278536
Adipate	0.13815294	0.19626967	-0.2545485	0.32142918
Lactate	0.17015011	0.01694167	-0.2926651	0.29315501
Hypoxanthine	0.03908591	0.15958037	-0.2230033	0.2742196
Glutamate	0.17889476	0.13997017	-0.2089776	0.25152192
Alanine	0.20568069	-0.2205282	-0.1199601	0.25104403
Acetone	0.10609781	0.2141336	-0.1285256	0.24974391
Phenylalanine	0.15497467	0.0416933	0.23759456	0.24122501
O-phosphocholine	0.13681365	0.16962719	-0.1660462	0.23737041
Leucine	0.2067085	0.10994236	0.20055647	0.22871427
3-Methyl-2-Oxovalerate	0.12816517	0.21837494	0.04288388	0.22254582
Creatine Phosphate	0.15903242	-0.2070038	-0.0426857	0.21135904
Histidine	0.01717568	0.0594115	0.20228903	0.21083305
3-Aminoisobutyrate	0.10503272	0.19351667	0.02292365	0.19486969
Valine	0.22254027	0.01077726	0.17203364	0.17237089
Pyruvate	0.1708877	0.01863289	-0.166928	0.16796474
Choline	0.19856831	-0.0974704	0.12782132	0.1607444
Succinate	0.140404	0.01503279	-0.1588119	0.15952176
Creatine	0.15406358	0.05461516	0.14575958	0.15565562
Formate	0.12481118	0.05078229	-0.1446407	0.15329638
Malonate	0.16357974	-0.1040489	0.11232368	0.15311039
Betaine	0.18953171	-0.0917461	0.11954568	0.15069343
Aspartate	0.12217261	-0.0824399	-0.1216538	0.1469557
Proline	0.19982237	-0.1452455	0.00767188	0.14544797
Glutamine	0.21257371	-0.1182284	0.07247271	0.13867315
Acetate	0.13445786	0.13279824	-0.038851	0.13836464
Isoleucine	0.18346117	0.02099238	0.13570294	0.13731703
Lysine	0.23059301	-0.1130716	0.06653587	0.1311953
Asparagine	0.16265794	-0.1146923	-0.0516647	0.12579175
Methionine	0.19368617	-0.1028348	0.07213219	0.12561072
Tyramine	0.1557889	-0.0037425	-0.1187582	0.11881712
2-Oxovalerate	0.15110179	0.08529709	0.03026353	0.09050676
Dimethylamine	0.12591453	-0.0805601	0.03061979	0.08618293
Creatinine	0.20879239	-0.0609296	0.03552048	0.07052746
Sn-Glycero-3-phosphocholine	0.18485381	-0.0376795	-0.0318023	0.0493065
Threonine	0.14074561	0.01897372	-0.0422133	0.04628141
Glucose	0.19468583	-0.0203949	0.00310279	0.02062958

PC2 and PC3 show a clear separation between samples from trauma patients and samples from other research participants. The table is sorted according to the magnitude of the loadings vectors in PC2 and PC3 (column 4). Magnitude of PC2 and PC3 was calculated as follows: $[(PC2\ Loading)^2 + (PC3\ Loading)^2]^{(1/2)}$
PC2, principal component 2; PC3, principal component 3; TMAO, trimethylamine N-oxide.

Table 2.
Principal component loadings for the first three components for all profiled metabolites.

studies, we performed a meta-analysis of the metabolic profiles obtained from these previously published studies [12–15]. Our results suggest that samples from trauma patients are distinguishable from healthy controls and patients with respiratory failure or acute pancreatitis. Principal components 2 and 3 can be used to separate trauma patients' samples from other samples, and highlight oxidative stress and disruptions to energy metabolism.

Traumatic injury is known to have a profound effect on molecular processes, impacting more than 80% of cellular functions and pathways, earning the moniker "genomic storm" [18]. In light of this, it is unsurprising that our unsupervised analysis would separate trauma samples from non-trauma samples. In our own work evaluating metabolomes of trauma patients age 21–40 years and trauma patients older than 65 years, we found a clear difference between metabolic profiles of younger healthy controls and older healthy controls. However, the data forced us to reject our hypothesis that metabolomes of older trauma patients would be distinguishable from younger trauma patients [13]. One interpretation of these data is that trauma deals a massive insult to metabolism that completely overtakes any baseline differences in metabolism caused by age.

Trauma from unintentional injury is the most common cause of death for persons age 44 and under [19]. Treatment of traumatic injury remains limited to supportive care such as stopping any bleeding and giving fluids to resuscitate. Lacking specific therapies for traumatic injury, early treatment is a key to improving survival. Metabolomics has already been successfully used to identify succinate, an objective biomarker of mortality, to improve triage [20–22]. However, new technology needs to be developed to bring succinate detection and quantification to the clinic.

5. Improving patient monitoring with metabolomics

It may be surprising that NMR, with its relatively low resolution, can discriminate metabolic profiles of trauma patients from others. However, this technique does not reflect the extremely variable, highly individualized nature of critical illness. Improving the sensitivity of metabolite detection with mass spectrometry is required to highlight these features of critical illness.

In a preliminary study (manuscript in preparation), we used mass spectrometry to generate metabolic profiles of five ICU patients and five healthy controls. Samples were collected every 4 h for a period of 24 h. A standard methanol/acetone protocol was used to extract metabolites. A Q Exactive™ Quadrupole-Orbitrap mass spectrometer (Thermo Fisher Scientific, Waltham, MA) was employed for mass analysis. Analysis was performed in positive mode over a mass range of 70–1050 m/z. Spectra were aligned and processed with Progenesis QI software (Nonlinear Dynamics, Durham, NC).

Spectral intensities of 15,000 features identified by Progenesis QI were log-transformed and principal components analysis was performed using R software. The resulting scores are shown in **Figure 6**. Scores were colored by participant. A single, relatively tight cluster of healthy controls is clearly visible in the upper left quadrant of **Figure 6** (HC01-HC05, colored blue, green, and pink). Strikingly, each ICU patient (ICU01-IC05, red, orange, and purple) is clearly visible, and each patient forms its own unique cluster. Interestingly, the ICU patient colored in red was demonstrably less sick than the other patients, with a lower APACHE II (acute physiology and chronic health evaluation) score and a shorter ICU length of stay. It is likely that the sampling frequency combined with the sensitivity of mass spectrometry allowed us to see such highly individualized patterns in the metabolic profiles.

Based on these data, we posit that mass spectrometry-based metabolomics offers a unique way to characterize the highly individual, highly variable nature of critical illness. The PCA scores in **Figure 6** further offer the tantalizing suggestion that metabolic profiles reflect severity of illness, since the scores of patient with the lower APACHE II score and shorter ICU stay were closest to the scores of the healthy controls. We further hypothesize that, tracked over time, principal component analysis of individual patients' metabolic profiles could offer insight

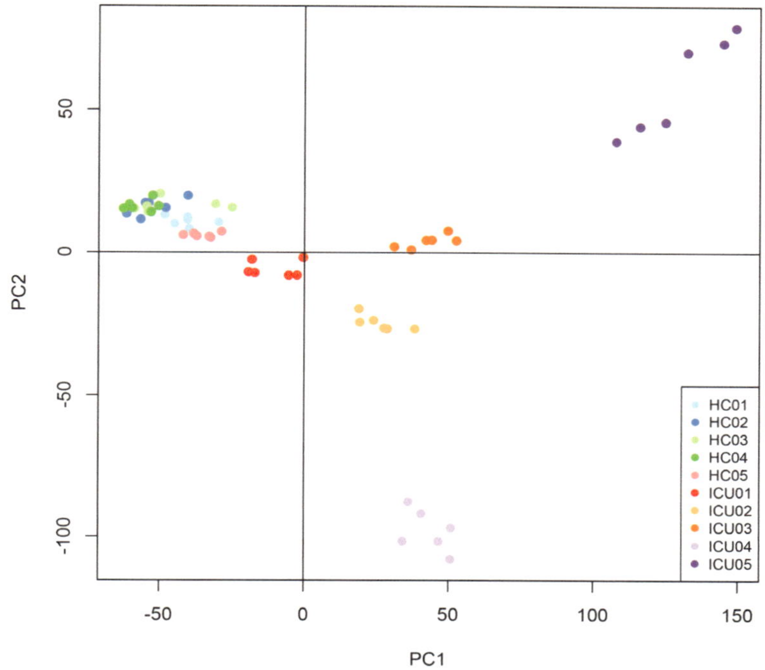

Figure 6.
Scores plot of the first two principal components, constructed from metabolic profiles of five ICU patients and five healthy controls. Samples are colored by individual study subject. Subjects HC01 through HC05 (top left quadrant) are healthy controls. Subjects ICU01-ICU05 are ICU patients. Samples were collected in each participant every 4 h for 24 h. PCA clearly illustrates the extreme variability and individuality of metabolic profiles in critical illness.

into their clinical courses, moving farther away from a "healthy" profile as conditions worsen and closer to a "healthy" profile as conditions improve. Others have used principal component analysis of circulating inflammatory cytokines in a similar way, to identify individual patients who go on to develop multiple organ dysfunction syndrome [23].

This line of inquiry is not without challenges. Collecting samples frequently around the clock from human study participants is a challenge, and a substantial sample bank will have to be obtained to establish a "healthy" metabolic profile. Mass spectrometry results in a large, extremely rich data set of features which are difficult to map to individual metabolites, so it is difficult to identify the set of metabolites that drive the patterns observed here.

Finally, patients will have to be monitored over time and their metabolic profiles will have to be mapped to their outcomes in order to link their "trajectories" to outcomes or adverse events. This may be a daunting task. However, others have successfully established continuous predictive analytics monitoring from physiologic data in neonatal ICUs [24]. Continuous predictive analytics monitoring allows ICU staff to follow patient trajectories that serve as an early-warning system for sepsis [25], allowing for earlier treatment before inflammation and infection worsen. Since, as with trauma, early intervention is a key to survival from sepsis, bringing predictive monitoring to the ICU is a clear way to improve patient outcomes.

New technology needs to be developed to bring metabolomics to the bedside if it is to be used to track patient trajectories in a clinically useful manner. In the meantime, much can be learned about critical illness from metabolomics.

6. Conclusions

Critical illness encompasses a variety of life-threatening conditions characterized by the need for frequent, intensive interventions. Patients are heterogeneous and may not respond to treatments in a predictable way; further, their conditions can change quickly over time. Metabolomics, reflective of the phenome, has great potential to impact patient care. NMR-based metabolomics highlights trauma as having a unique impact on the metabolome relative to healthy controls and other conditions. Mass spectrometry, with its increased sensitivity over NMR, highlights an extremely individualized variation in the metabolomes of ICU patients that does not exist in healthy controls. With technological innovations to bring metabolomics to the bedside, it may be used in the future to bring predictive analytics to the ICU, leading to faster and more appropriately individualized interventions, and improving patient care and outcomes.

Acknowledgements

Dr. Lusczek would like to acknowledge Dr. Greg Beilman and Dr. Sayeed Ikramuddin of the University of Minnesota, Department of Surgery. NMR instrumentation was provided by the Minnesota NMR Center. Funding for NMR instrumentation was provided by the Office of the Vice President for Research, the Medical School, the College of Biological Science, NIH, NSF, and the Minnesota Medical Foundation. Mass spectrometry metabolomics data were obtained by the University of Minnesota's Center for Mass Spectrometry and Proteomics.

Conflict of interest

Dr. Lusczek is on the board of directors of the Society for Complex Acute Illness.

Author details

Elizabeth R. Lusczek
Department of Surgery, University of Minnesota, Minneapolis, MN, USA

*Address all correspondence to: lusc0006@umn.edu

References

[1] Coopersmith CM, Wunsch H, Fink MP, Linde-Zwirble WT, Olsen KM, Sommers MS, et al. A comparison of critical care research funding and the financial burden of critical illness in the United States. Critical Care Medicine. 2012;**40**(4):1072-1079

[2] Rawal G, Yadav S, Kumar R. Post-intensive care syndrome: An overview. Journal of Translational Internal Medicine. 2017;**5**(2):90-92

[3] Elliott D, Davidson JE, Harvey MA, Bemis-Dougherty A, Hopkins RO, Iwashyna TJ, et al. Exploring the scope of post-intensive care syndrome therapy and care: Engagement of non-critical care providers and survivors in a second stakeholders meeting. Critical Care Medicine. 2014;**42**(12):2518-2526

[4] Hoffman LA. Post intensive care syndrome: Risk factors and prevention strategies. Critical Care Alert. 2015;**22**(12):89-93

[5] Society of Critical Care Medicine. Critical Care Statistics [Internet]. Available from: https://www.sccm.org/Communications/Critical-Care-Statistics [Accessed: 27 March 2019]

[6] Sweeney TE, Khatri P. Generalizable biomarkers in critical care: Toward precision medicine. Critical Care Medicine. 2017;**45**(6):934

[7] Maslove DM, Lamontagne F, Marshall JC, Heyland DK. A path to precision in the ICU. Critical Care. 2017;**21**(1):79

[8] Seymour CW, Gomez H, Chang C-CH, Clermont G, Kellum JA, Kennedy J, et al. Precision medicine for all? Challenges and opportunities for a precision medicine approach to critical illness. Critical Care. 2017;**21**(1):257

[9] Schmerler D, Neugebauer S, Ludewig K, Bremer-Streck S, Brunkhorst FM, Kiehntopf M. Targeted metabolomics for discrimination of systemic inflammatory disorders in critically ill patients. Journal of Lipid Research. 2012;**53**(7):1369-1375

[10] Antcliffe D, Gordon AC. Metabonomics and intensive care. Critical Care. 2016;**20**(1):68

[11] Rogers AJ, McGeachie M, Baron RM, Gazourian L, Haspel JA, Nakahira K, et al. Metabolomic derangements are associated with mortality in critically ill adult patients. PLoS One. 2014;**9**(1):e87538

[12] Lusczek ER, Muratore SL, Dubick MA, Beilman GJ. Assessment of key plasma metabolites in combat casualties. Journal of Trauma and Acute Care Surgery. 2017;**82**(2):309-316

[13] Lusczek ER, Myers C, Popovsky K, Mulier K, Beilman G, Sawyer R. Plasma metabolomics pilot study suggests age and sex-based differences in the metabolic response to traumatic injury. Injury. 2018;**49**(12):2178-2185

[14] Lusczek ER, Colling K, Muratore S, Conwell D, Freeman M, Beilman G. Stereotypical metabolic response to endoscopic retrograde cholangiopancreatography show alterations in pancreatic function regardless of post-procedure pancreatitis. Clinical and Translational Gastroenterology. 2016;**7**(5):e169

[15] Fortis S, Lusczek ER, Weinert CR, Beilman GJ. Metabolomics in COPD acute respiratory failure requiring noninvasive positive pressure ventilation. Canadian Respiratory Journal. 2017;**2017**:9480346. DOI: 10.1155/2017/9480346. 9pp

[16] Weljie AM, Newton J, Mercier P, Carlson E, Slupsky CM. Targeted profiling: Quantitative analysis of ^{1}H

NMR metabolomics data. Analytical Chemistry. 2006;**78**(13):4430-4442

[17] R Development Core Team. R: A Language and Environment for Statistical Computing. Vienna, Austria: R Foundation for Statistical Computing. 2018. Available from: http://www.R-project.org/

[18] Xiao W, Mindrinos MN, Seok J, Cuschieri J, Cuenca AG, Gao H, et al. A genomic storm in critically injured humans. The Journal of Experimental Medicine. 2011;**208**(13):2581-2590

[19] National Center for Injury Prevention and Control (NCIPC). Web-based Injury Statistics Query and Reporting System. Atlanta, GA: Centers for Disease Control and Prevention. Available from: https://webappa.cdc.gov/sasweb/ncipc/leadcause.html

[20] Witowski NE, Lusczek ER, Determan CE, Lexcen DR, Mulier KE, Wolf A, et al. Metabolomic analysis of survival in carbohydrate pre-fed pigs subjected to shock and polytrauma. Molecular BioSystems. 2016;**12**(5):1638-1652

[21] D'alessandro A, Moore HB, Moore EE, Reisz JA, Wither MJ, Ghasasbyan A, et al. Plasma succinate is a predictor of mortality in critically injured patients. Journal of Trauma and Acute Care Surgery. 2017;**83**(3):491-495

[22] Lexcen DR, Lusczek ER, Witowski NE, Mulier KE, Beilman GJ. Metabolomics classifies phase of care and identifies risk for mortality in a porcine model of multiple injuries and hemorrhagic shock. The Journal of Trauma and Acute Care Surgery. 2012;**73**(2):S147-SS55

[23] Namas RA, Almahmoud K, Mi Q, Ghuma A, Namas R, Zaaqoq A, et al. Individual-specific principal component analysis of circulating inflammatory mediators predicts early organ dysfunction in trauma patients. Journal of Critical Care. 2016;**36**:146-153

[24] Moss TJ, Lake DE, Calland JF, Enfield KB, Delos JB, Fairchild KD, et al. Signatures of subacute potentially catastrophic illness in the intensive care unit: Model development and validation. Critical Care Medicine. 2016;**44**(9):1639

[25] Keim-Malpass J, Kitzmiller RR, Skeeles-Worley A, Lindberg C, Clark MT, Tai R, et al. Advancing continuous predictive analytics monitoring: Moving from implementation to clinical action in a learning health system. Critical Care Nursing Clinics of North America. 2018;**30**(2):273-287

Chapter 5

Chinese Medicines for Cancer Treatment from the Metabolomics Perspective

Wei Guo, Hor-Yue Tan, Ning Wang and Yibin Feng

Abstract

Cancer is one of the most prevalent diseases all over the world with poor prognosis and the development of novel therapeutic strategies is still urgently needed. The large amount of successful experiences in fighting against cancer-like diseases with Chinese medicine has suggested it as a great source of alternative treatments to human cancers. Cancer cells have been shown to own a predominantly unique metabolic phenotype to facilitate their rapid proliferation. Metabolic reprogramming is a remarkable hallmark of cancer and therapies targeting cancer metabolism can be highly specific and effective. Based on the sophisticated study of small molecule metabolites, metabolomics can provide us valuable information on dynamically metabolic responses of living systems to certain environmental condition. In this chapter, we systematically reviewed recent studies on metabolism-targeting anticancer therapies based on metabolomics in terms of glucose, lipid, amino acid, and nucleotide metabolisms and other altered metabolisms, with special emphasis on the potential of metabolic treatment with pure compounds, herb extracts, and formulations from Chinese medicines. The trends of future development of metabolism-targeting anticancer therapies were also discussed. Overall, the elucidation of the underlying molecular mechanism of metabolism-targeting pharmacologic therapies will provide us a new insight to develop novel therapeutics for cancer treatment.

Keywords: metabolomics, cancer metabolism, adjuvant therapies, Chinese medicines

1. Introduction

Despite all recent improvements in early detection and pleiotropic therapeutics, cancer is still the leading cause of death all over the world [1]. It is one of the most prevalent diseases with complex risk factors, and the mortality rate is similar to its morbidity, which reflects its poor prognosis. It has been projected that approximately 3.12 million new cases of cancer and a cancer death toll of 2.5 million will occur per year in China, which brings a huge burden on society [2]. To date, there are three conventional cancer therapies for cancer, including surgical resection, chemotherapy, and radiotherapy. However, diverse drawbacks and limitations have been observed in these cancer therapies either alone or in combination. For example, most cancer patients are not suitable to undergo the surgical resection due to the late diagnosis and other factors. As the major therapies for cancer patients in middle and advanced stages, chemotherapy and radiotherapy have been shown to present serious side effects and complications, such as myelosuppression, hematological toxicity, cardiac damage, and liver and kidney dysfunction [1, 3]. Moreover, tumor cells have the ability to develop resistance to evade cell death, and the therapeutic efficacy of the current chemotherapeutic drugs is significantly reduced by the increasingly acquired drug resistance [4]. Therefore, it imminently deserves to develop more effective and less toxic adjuvant therapies for cancer prevention and treatment.

1.1 Cancer metabolism

It has been reported that cell metabolism has an essential role in the pathological progression of cancer and metabolic reprogramming is a remarkable hallmark of cancer [5]. Cancer cells have been shown to own a predominantly unique metabolic phenotype to facilitate their rapid proliferation, which is dramatically different from normal cells. Cancer cells tend to acquire energy via glycolysis rather than the much more efficient oxidative phosphorylation pathway even in aerobic conditions, which is the famous phenomenon of cancer called the "Warburg effect" [6]. Besides the consumption of glucose, cancer cells have also been reported to favor glutamine as a preferential fuel [7]. Accumulating evidences indicate that mutations in metabolic enzymes can promote the development of cancer. For example, mutations in the tricarboxylic acid (TCA) cycle enzyme isocitrate dehydrogenase, succinate dehydrogenase, and fumarate hydratase can affect the corresponding metabolites a-ketoglutarate, succinate, and fumarate. These changes can further affect the 2-oxoglutarate-dependent dioxygenases and then result in some cancers, such as paraganglioma and renal cell cancer [8–10]. What is more, the drug resistance of cancer cells is also shown to be associated with their metabolic alterations [11]. In this perspective, cancer metabolism has become a potentially fertile area, and therapies targeting cancer metabolism can be highly specific and effective. Nowadays metabolism-targeting anticancer therapies are drawing researchers' great attention and becoming a new therapeutics for cancer treatment [12].

1.2 Metabolomics and cancer

As a valuable complement to emerging "omics" science including genomics, transcriptomics, and proteomics, metabolomics utilizes leading-edge analytical chemistry technologies and advanced computational approaches to characterize the small endogenous and exogenous molecule metabolites in various biochemical metabolisms from complex biochemical mixtures [13]. Metabolomics can provide us a direct readout on dynamically metabolic responses of living systems to certain genetic modifications or pathophysiological stimuli [14], which has been extensively adopted in the field of disease diagnosis, pharmacodynamic evaluation, therapeutical monitoring, and drug discovery [15]. There are three main analytical chemistry platforms in metabolomics research, namely, nuclear magnetic resonance (NMR) spectroscopy, liquid chromatography mass spectrometry (LC-MS), and gas chromatography MS (GC/MS). Each platform has its own strengths and limitations. There are three main methodological approaches to analyze the small metabolites in metabolomics, namely, targeted, untargeted, and stable isotope-resolved metabolomics (SIRM). Numerous systemic reviews have shown in detail how each analytical platform and methodological approach works in metabolomics studies [16–20].

As mentioned above, cell metabolism has an essential role in the pathological progression of cancer, and metabolic reprogramming is a remarkable hallmark of cancer. In this context, it would be conducive to employ metabolomics in the field of cancer research for exploration of tumorigenesis mechanisms, diagnosis and monitoring of tumor, as well as discovery of novel anticancer therapies [21–23].

1.3 Chinese medicines and cancer treatment

Due to their various biological activities and low toxicity, natural products derived from Chinese medicines are reported to be an excellent source for anticancer drugs as a complementary and alternative approach [24]. Chinese medicines have evolved with their own unique theoretical system in Asian countries, especially China over thousands of years. Chinese medicines are usually divided into pure compounds, herb extracts, and formulations. Formulations from Chinese medicines are extensively employed in Chinese hospitals for clinical cancer treatment [25]. Numerous Chinese herb extracts have been reported to show inhibitory effects on cancers [26]. An increasing number of pure compounds derived from Chinese medicine herbs have been shown to inhibit the development of cancers through various mechanisms [27–30]. Besides, a large number of studies have revealed that Chinese medicines in combination with conventional chemotherapy and radiotherapy could increase the therapeutic efficacy and decrease the serious side effects and complications of these therapies [31, 32]. It is convinced that Chinese medicines are gaining increasing reputation and credibility as adjuvant therapies for cancer prevention and treatment.

Although Chinese medicines have been employed in cancer prevention and treatment for a long time, the underlying mechanisms on how they work remain to be fully elucidated because of their unique medical system with multicomponent nature. In accordance with the holistic perspective of Chinese medicines, metabolomics opens up a unique and novel insight into efficacy evaluation and action mechanism exploration of Chinese medicines as adjuvant therapies for cancer prevention and treatment.

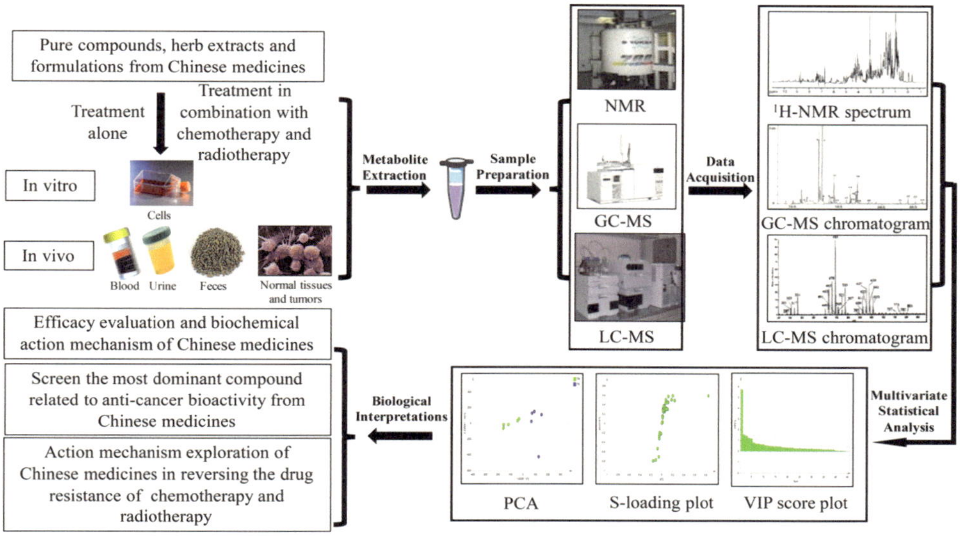

Figure 1.
The typical flowchart of metabolomics studies on antineoplastic Chinese medicines.

References	Pure compound	Cancer	Study	Method	Significantly changed metabolites or pathways	Main findings
26,700,591	Geranylgeranoic acid	Hepatoma	HuH-7 cells in vitro	UPLC/TOF/MS	GGA induced a time-dependent increase in the cellular contents of fructose 6-phosphate and decrease of fructose 1,6-diphosphate	GGA may shift HuH-7 cells from aerobic glycolysis to mitochondrial respiration through the immediate upregulation of TIGAR and SCO2 protein levels
26,160,839	Halofuginone	Colorectal cancer	HCT116 cells in vitro	UPLC-MS/MS, GC/MS and UPLC/LTQ-Orbitrap MS	Metabolomics delineated the slower rates in both glycolytic flux and glucose-derived tricarboxylic acid cycle flux	HF regulates Akt/mTORC1 signaling pathway to dampen glucose uptake and glycolysis in CRC cells
29,589,762	(−)-5-Hydroxy-equol	Hepatocellular Carcinoma	SMMC-7721 cells in vitro	1H NMR	(−)-5-Hydroxy-equol treatment significantly altered energy and amino acid metabolism	Integrated metabolomics and further verifications may facilitate the exploration of the anti-HCC mechanisms of (−)-5-hydroxy-equol
29,802,724	Nummularic acid (NA)	Prostate cancer (PCa)	DU-145 and C4-2 cells in vitro	ALEX-CIS GC–TOF-MS	The metabolism pathways related to glycolysis, TCA, and glutamine metabolisms were changed after NA treatment	NA may induce energy crisis to inhibit PCa
30,391,728	Magnoline	Prostate cancer	22RV1 cells in vitro	UPLC-MS	Magnoline markedly restored the energy metabolism, amino acid metabolism, and fatty acid metabolism	Cancer cells may result in death because of insufficient material basis to favor their rapid proliferation

References	Pure compound	Cancer	Study	Method	Significantly changed metabolites or pathways	Main findings
28,651,973	1,25-Dihydroxyvitamin D3	Prostate cancer	LNCaP cells in vitro	GC/ MS	1,25(OH)2D3 decreased glucose uptake and increased citrate/isocitrate due to TCA cycle truncation	Re-wiring glucose metabolizing pathways, and induction of a “differentiated” metabolic phenotype by 1,25(OH)2D3, may prove clinically beneficial
26,541,605	Vitamin C	Colorectal cancer	KRAS and BRAF mutant lines and isogenic wild-type counterparts in vitro	LC-MS/MS	High levels of vitamin C increased uptake of dehydroascorbic acid (DHA) and decreased glutathione	These results provide a mechanistic rationale for exploring the therapeutic use of vitamin C for CRCs with KRAS or BRAF mutations
28,916,726	β-Lapachone	Pancreatic ductal adenocarcinoma	MiaPaCa2 cells in vitro	GC/MS and 1H NMR	β-lap treatment was found to decrease the NAD-sensitive pathways, such as glycolysis and TCA cycle	Targeting NQO1 may sensitize the treatment of β-lap
28,737,429	Diethylstilbestrol	Prostate cancer	PC3 cells in vitro	1H NMR	Lactate, phosphocreatine, and GSH were the biomarkers for DES treatment	DES upon conjugation had a more specific effect and less toxicity
28,918,937	Koningic acid	Colorectal cancer	HCT116 cells in vitro	Integrated pharmacogenomics and LC-HRMS metabolomics	Glycolysis was the highest scoring pathway only in KA-treated cells	KA efficacy is not determined by the status of individual genes but by the quantitative extent of the WE, leading to a therapeutic window in vivo

References	Pure compound	Cancer	Study	Method	Significantly changed metabolites or pathways	Main findings
30,114,709	Omega-3 polyunsaturated fatty acids	Breast cancer	MCF7 cells in vitro	GC/MS	Glycolysis and glutamine metabolism pathways were markedly reduced when treated with the combination of Rp and ω-3 PUFAs	ω-3 PUFA could increase the anti-breast cancer potential of Rp
28,198,625	Curcumin	Hepatocarcinoma	Serum from DEN-induced hepatocarcinogenesis model	GC/MS	Curcumin attenuated metabolic disorders via increasing concentration of glucose and fructose, and decreasing levels of glycine and proline	Curcumin exhibited a potent liver protective agent inhibiting chemically induced liver injury through suppressing liver cellular metabolism in the prospective application
29,448,205	6,7-Dimethoxy-1,2,3,4-tetrahydro-iso- quinoline-3-carboxylic acid	Colorectal carcinoma	Serum from DMH-induced CRC albino Wistar rat model in vivo	1H NMR	M1 exhibited to reverse the perturbed metabolism pathways in CRC condition, including glycolysis, TCA cycle, choline, phosphatidylinositol and gluconeogenesismetabolisms	M1 has the anti-CRC potential via the blockade of IL-6/JAK2/STAT3 oncogenic signaling
27,416,811	Physapubenolide	Hepatocellular carcinoma	HepG2 cells in vitro and tumor tissues and plasma from a mouse-xenograft model bearing liver carcinoma H22 cells in vivo	GC/MS	PB disturbed the metabolic pattern and significantly decreased lactate production	PB exhibits anticancer activities through suppression of glycolysis via the Akt-p53 pathway
30,322,263	Naringenin	Lung cancer	Serum from the urethane-induced lung cancer rat model in vivo	1H NMR	The glycolysis was restored to normal levels with co-therapy of Gnb and Nar	Co-therapy has the superiority over alone treatment to improve the therapeutic efficacy

References	Pure compound	Cancer	Study	Method	Significantly changed metabolites or pathways	Main findings
26,859,520	Flexibilide	Colon cancer	HCT-116 cells in vitro	UPLC/Q-TOF MS	Flexibilide exhibited the therapeutic effect on colon cancer mainly via downregulating PC biosynthesis pathway	Flexibilide exhibited the therapeutic effect on colon cancer mainly via down-regulating PC biosynthesis pathway
28,296,891	Englerin A	Clear cell renal carcinoma	A498 cells in vitro	LC-MS/MS	Englerin A significantly reversed lipid metabolism and increase ceramides levels	Ceramides may be a mediator of some of the actions of englerin A
28,948,276	Isoquercitrin	Bladder cancer	T24 cells in vitro	UPLC/Q-TOF MS	Isoquercitrin treatment was found to regulate lipid and anaerobic glycolysis	ISO influenced T24 bladder cancer cell metabolism, and this process was mainly involved in activating the AMPK pathway
28,496,003	Peiminine	Colorectal cancer	UPLC-MS and GC/MS	UPLC-MS and GC/MS	Peiminine treatment altered several metabolites, including lignocerate (24:0), oleate (18:1n9), glutamine, and glucose	Peiminine exerted the predominant therapeutic effect mainly via the metabolic regulation of lipids, amino acids, and carbohydrates
29,321,577	8u	Hepatocellular carcinoma	HepG2 cells in vitro	UPLC/Q-TOF MS	8u was found to significantly inhibit the invasion and metastasis of HepG2 cells and regulate intracellular lipid metabolism	8u could efficiently suppress the invasion and metastasis of HepG2 cells by decreasing the expression of HSP90α protein and inhibiting the PI3K/Akt signaling pathway

References	Pure compound	Cancer	Study	Method	Significantly changed metabolites or pathways	Main findings
28,125,641	Genistein and calcitriol	Osteosarcoma	MG-63 Cells in vitro	GC/MS	Co-therapy of genistein and calcitriol was found to regulate lipids and amino acids rather than energy metabolism	The promotional effects of high level of genistein on osteosarcoma could be decreased by the co-treatment of calcitriol
27,533,043	Silibinin	Prostate cancer	Tumor tissues from 22Rv1 Xenograft model in vivo	1H-NMR	Silibinin treatment did not greatly affect glucose uptake of PCa tumor but decreased the lipid synthesis	These findings further support silibinin usefulness against PCa through inhibiting hypoxia-induced signaling
26,744,170	Acyclic retinoid	Hepatocellular carcinoma	Liver tissues from mouse DEN-induced HCC model in vivo	CE-TOFMS and LC-TOFMS	ACR predominantly reversed lipogenesis but not glucose metabolism by inhibiting linoleic acid metabolites	Lipid metabolic reprogramming plays a critical role in the protective effects of ACR on HCC
30,871,192	Delta-tocotrienol	Non-small cell lung cancer	A549 and H1299 cells in vitro	1H-NMR	Cellular metabolomics analysis showed significant inhibition in the uptake of glutamine, its derivatives glutamate and glutathione, and some EAAs in both cell lines with δT treatment	δT treatment could suppress the glutamine uptake via suppressing glutamine transporters and then resulted in the induction of apoptosis and suppression of cell proliferation

References	Pure compound	Cancer	Study	Method	Significantly changed metabolites or pathways	Main findings
30,068,874	Celastrol	Colon cancer	HCT116 cells in vitro	UPLC/MS	Metabolomics analysis found celastrol changed the levels of lipid markers, carnitine and amino acids. Tryptophan was further identified as a special biomarker by targeted metabolite analysis	The suppression of IDO expression and tryptophan catabolism may be part of the mechanisms of celastrol in its cytotoxic effect against HCT116 colon cancer cells
27,754,384	Melittin	Ovarian cancer	A2780 and A2780CR cell lines in vitro	LC-MS	Melittin treatment of cisplatin-sensitive cells decreased glutamine, proline, and arginine pathways	Melittin might have some potential as an adjuvant therapy in cancer treatment
28,674,386	Chlorogenic acid and caffeic acid	Hepatocellular carcinoma	Serum from DEN-induced HCC model in vivo	16 S rRNA and LC-MS, GC/MS-MS, GC/MS	Both CaA and ChA treatment reverse 28 metabolites	The levels of ethanolamine, L-methionine, L-tyrosine, and bilirubin were associated with diminished Prevotella 9 and Lachnospiraceae incertae sedis and elevated Ruminococcaceae UCG-004
29,202,102	Resveratrol, curcumin and ursolic acid	Prostate cancer	Serum from a mouse allograft model of prostate cancer in vivo	LC-MS and GC/MS	Glutamine metabolism was regulated by the compound combinations	Compared with the individual treatment, the combined treatment has the greater antineoplastic property

References	Pure compound	Cancer	Study	Method	Significantly changed metabolites or pathways	Main findings
29,651,531		Hepatoma	SMMC7721 cells in vitro	GC/MS and LC/MS	GLA treatment diminished amino acid metabolism and elevated the metabolisms of sphingolipid, purine, and pyrimidine	GC/MS- and LC/MS-based metabolomics applied to cell culture enhanced our current understanding of the metabolic response to GLA treatment and its mechanism
26,851,007	Taurine	Breast cancer	Plasma from dimethylbenz[a]anthracene-induced breast carcinogenesis in rats in vivo	GC–TOFMS	Taurine treatment regulated 23 differential metabolites, which were associated with glucose, energy and amino acid, as well as nucleic acid metabolism	The antitumor activity of taurine in rats is mediated through altered metabolism of breast cancer cells
27,374,097	Celastrol	Acute promyelocytic leukemia	HL-60 cells in vitro and tumor tissue from mice xenograft model in vivo	UPLC-MS	Celastrol treatment regulated uridine metabolite, which further enhances apoptosis	The study firstly reveals that uridine deficiency contributes to mitochondrial apoptosis induced by celastrol in APL cells
29,787,425	Gamma-tocotrienol	Cancer	Serum from nonhuman primate models in vivo	UPLC/Q-TOF MS	GT3 could regulate the changed fatty acid beta-oxidation, amino acid and purine catabolism metabolism caused by irradiation	This initial assessment also highlights the utility of metabolomics in determining underlying physiological mechanisms responsible for the radioprotective efficacy of gamma-tocotrienol

References	Pure compound	Cancer	Study	Method	Significantly changed metabolites or pathways	Main findings
27,335,141	Fisetin	Prostate cancer	Tumor tissues from prostate cancer xenografts in vivo	HPLC/ESI–MS	Fisetin treatment was shown to downregulate secreted and intracellular hyaluronan (HA), which conferred resistance to prostate oncogenesis	Fisetin is an effective, nontoxic, potent HA synthesis inhibitor
29,978,476	Galactolipid 1,2-di-O-linolenoyl-3-O- β-galactopyranosyl-sn-glycerol	Melanoma	Serum from a syngeneic mouse model implanted with B16 melanoma in vivo	LC-MS/MS	dLGG treatment markedly elevated 12/15-LOX catalyzed oxylipin products in serum	This study shows the novel therapeutic effect of phytoagent dLGG and suggests its potential as a therapeutic agent for metastatic melanoma
30,668,340	Deoxyelephantopin	Melanoma	Kidney tissues from murine B16 metastatic allograft model in vivo	UPLC/ESI-QTOF MS	Co-therapy of DET and cisplatin could reverse the changed urea cycle metabolites and hippuric acid in renal tissues caused by cisplatin	The co-therapy of DET and cisplatin could be an effective treatment with low toxicity for melanoma

Table 1.
Summary of recent metabolomic studies on anticancer therapies of pure compounds from Chinese medicines.

Increasing excellent reviews have been focused on the application of metabolomics in the metabolic changes and the possible underlying mechanisms behind these alterations in the pathogenesis of different kinds of cancer [33–35]. Little reviews have been highlighted on the metabolism-based anticancer therapies. Since Chinese medicine has been suggested to be a great source of alternative treatments to human cancers, in this chapter we systematically reviewed recent studies from 2015 to March 2019 on metabolism-targeting anticancer therapies based on metabolomics in terms of glucose, lipid, amino acid, and nucleotide metabolisms and other altered metabolisms, with special emphasis on the potential of metabolic treatment with pure compounds, herb extracts, and formulations from Chinese medicines. The typical flowchart of metabolomics studies on antineoplastic Chinese medicines is shown in **Figure 1**. **Table 1** summarized the recent metabolomics studies on anticancer therapies of pure compounds from Chinese medicines. At the same time, the trends of future development of metabolism-targeting anticancer therapies were also discussed.

2. Review on metabolism-targeting Chinese medicine treatment on human cancers

2.1 Glucose metabolism

As mentioned above, cancer cells tend to acquire energy via glycolysis rather than the much more efficient oxidative phosphorylation pathway even in aerobic conditions. Glucose and energy metabolisms play an important role in the tumorigenesis of cancer and could be the therapeutic targets for cancer treatment. Pure compounds, herb extracts, and formulations from Chinese medicines, which target glucose and energy metabolisms, are attracting increasing attention for the development of anticancer therapies.

Geranylgeranoic acid (GGA), a kind of acyclic diterpenoids, is derived from some medicinal herbs such as turmeric. UPLC/TOF/MS-based metabolomics analysis was used to study the underlying anticancer mechanism of GGA in human hepatoma-derived HuH-7 cells [36]. It was found that GGA may shift the energetic state of HuH-7 cells from aerobic glycolysis to mitochondrial respiration, which was revealed by a time-dependent augment of fructose 6-phosphate and decline of fructose 1,6-diphosphate in HuH-7 cells after GGA treatment. Halofuginone (HF) is an active compound derived from the febrifugine which can be extracted from the Chinese herb *Dichroa febrifuga* Lour. Chen and his colleagues used the combination of UPLC-MS/MS, GC/MS, and UPLC/LTQ-Orbitrap MS metabolomics from HCT116 cells in vitro to study the anti-colorectal cancer (CRC) properties of HF [37]. They found the slower rates in the fluxes of both glycolytic and glucose-derived TCA cycle after HF treatment mainly via Akt/mTORC1 signaling pathway. (−)-5-Hydroxy-equol, as an isoflavone derived from microbial biotransformation, was shown to exhibit anti-hepatocellular carcinoma (HCC) potential. To explore the underlying mechanism, a ^{1}H NMR-based metabolomics of SMMC-7721 cells in vitro was conducted [38]. It was found that (−)-5-hydroxy-equol treatment significantly altered energy and amino acid metabolisms, which revealed that integrated metabolomics and further verifications may facilitate the exploration of the anti-HCC mechanisms of (−)-5-hydroxy-equol. Nummularic acid (NA) is a triterpenoid isolated from a medicinal plant *Fraxinus xanthoxyloides*. To explore its anticancer potential, a ALEX-CIS GC–TOF-MS-based metabolomics analysis of DU-145 and C4-2 cells in vitro was performed [39]. It was shown that the metabolism pathways related to glycolysis, TCA, and glutamine metabolisms were changed after NA treatment, which suggested NA may induce energy crisis to inhibit prostate cancer. Magnoline is the primary compound derived from *Cortex Phellodendri amurensis*, which exhibits significant therapeutic potential for PCa. Sun et al. conducted a UPLC-MS metabolomics of 22RV1 cells in vitro on PCa [40]. It was found that magnoline markedly restored the energy metabolism, amino acid metabolism, and fatty acid metabolism, which revealed that cancer cells may result in death because of insufficient material basis to favor their rapid proliferation. 1,25-Dihydroxyvitamin D3 (1,25(OH)2D3), also known as calcitriol, is one of the bioactive forms of nutraceutical vitamin D. Recently, its metabolism-modulating effects against PCa have been reported [41]. Based on the metabolomics analysis of LNCaP cells in vitro, 1,25(OH)2D3 inhibited glucose uptake and increased citrate/isocitrate because of TCA cycle truncation. The re-wiring glucose metabolizing pathways by 1,25(OH)2D3 may prove its metabolism-modulating effects against PCa. Yun et al. found that high exposed level of vitamin C could selectively kill CRC cells harboring KRAS or BRAF mutations [42]. In detail, based on the LC-MS/MS metabolomics between KRAS and BRAF mutant lines and isogenic wild-type counterparts in vitro, high level of exposure of vitamin C could increase uptake of dehydroascorbic acid by GLUT1 transporter and then decrease glutathione, which could inactivate glyceraldehyde 3-phosphate dehydrogenase (GAPDH). β-Lapachone (β-lap), as a quinone-containing compound derived from the *lapacho* tree located in South America, is bioactivated by NAD(P)H: quinone oxidoreductase 1 (NQO1). Recently, its effects on energy metabolism due to NAD^+ depletion on pancreatic ductal adenocarcinoma (PDA) have been shown [43]. Based on the combined GC/MS and ^1H NMR metabolomics analysis of MiaPaCa2 cells in vitro, β-lap treatment was found to decrease the NAD-sensitive pathways, such as glycolysis and TCA cycle, which revealed that targeting NQO1 may sensitize the treatment of β-lap. Diethylstilbestrol (DES), as a nonsteroidal estrogen, is the pharmacological inhibitor to HIF-1a. Arminan et al. employed NMR-based metabolomics of PC3 cells in vitro to explore the metabolic responses of PCa cells to hypoxia and the treatment of DES or its polyacetal conjugate tert-DES [44]. It was shown that lactate, phosphocreatine, and glutathione were the biomarkers for DES treatment. What is more, compared with tert-DES, the cell metabolome had a more extensive impact in the free DES treatment, which revealed that DES upon conjugation had a more specific effect and less toxicity. Koningic acid (KA), as an active natural product derived from the *Trichoderma fungus*, is a selective inhibitor of GAPDH. Recently Liberti et al. employed integrated pharmacogenomics and LC-HRMS metabolomics of HCT116 cells to explore the response of KA to CRC [45]. As a result, they found that partial GAPDH suppression is more selective for highly glycolytic tumors, underscoring the potential of targeting glucose metabolism therapy could be an integral part of precision medicine. Rapamycin (Rp) is widely used in the treatment of breast cancer. However, its efficacy has been significantly reduced by the increasing drug resistance and serious metabolic disorders. Dietary omega-3 polyunsaturated fatty acids (ω-3 PUFAs) have been reported to markedly inhibit breast cancer. To explore whether combined treatment of Rp and ω-3 PUFAs has better efficacy, a GC/MS-based metabolomics of MCF7 cells in vitro was done [46]. It was found that glycolysis and glutamine metabolism pathways were markedly reduced when treated with the combination of Rp and ω-3 PUFAs, suggesting that ω-3 PUFA could increase the anti-breast cancer potential of Rp. Curcumin, as the primary bioactive compound from the spice turmeric, was found to be a potent anticancer agent [47]. In detail, based on the serum metabolomics analysis, curcumin attenuated the metabolic disorders of diethylnitrosamine (DEN)-induced hepatocarcinogenesis by elevating the levels of glucose and fructose and reducing the levels of glycine and proline. 6,7-Dimethoxy-1,2,3,4-tetrahydro-isoquinoline-3-carboxylic acid (M1) is an isoquinoline alkaloid isolated from *Mucuna pruriens* seeds. To evaluate the anti-CRC effects of M1, ^{1}H NMR-based metabolomics of serum from dimethylhydrazine (DMH)-induced CRC albino Wistar rat model in vivo was conducted [48]. As a result, M1 exhibited to reverse the perturbed metabolism pathways in CRC condition, including glycolysis, TCA cycle, choline, and phosphatidylinositol and gluconeogenesis metabolisms. Taken together, this study offered that M1 had the anti-CRC potential via the blockade of IL-6/JAK2/STAT3 oncogenic signaling. Physapubenolide (PB) is a withanolide derived from *Physalis pubescens*. Recently its potential as a promising therapeutic drug has been put forward. However, the underlying mechanism of how it works remains to be explored. Ma et al. employed GC/MS-based metabolomics of both HepG2 cells in vitro and tumor tissues and plasma from a mouse-xenograft model bearing liver carcinoma H22 cells in vivo [49]. It was found that PB reversed the disturbed metabolic pattern by markedly decreasing the lactate production, suggesting PB may exhibit anti-HCC activities through suppression of glycolysis via the Akt-p53 pathway. Gefitinib (Gnb), as a tyrosine kinase inhibitor, is widely used for the treatment of lung cancer. However, the increasing drug resistance and serious metabolic disorders have significantly reduced its efficacy. Naringenin (Nar), as flavonoid isolated from citrus fruits, has been reported to show antioxidant, antimutagenic, and anticarcinogenic activities. To explore whether co-therapy through biotin-modified nanoparticles (NPs) of Gnb and Nar, a ^{1}H NMR-based metabolomics of serum from the urethane-induced lung cancer rat model in vivo was conducted [50]. It was found that the glycolysis was restored to normal levels with co-therapy of Gnb and Nar, which showed that co-therapy had the superiority over treatment only to improve the therapeutic efficacy.

Silymarin, extracted from the seeds of milk thistle (*Silybum marianum*), has the anti-inflammation activity. To explore the mechanism of how it suppresses inflammation, a combined transcriptional profiling and GC/MS metabolomics was conducted on Huh7-TLR3 cells [51]. It was found that the glycolytic, TCA cycle, and amino acid metabolism pathways were inhibited after silymarin treatment, which revealed that silymarin may have potential in defining how metabolic pathways mediate cellular inflammation. Rhizoma Paridis saponins (RPS) are the effective parts of Rhizoma Paridis, which have been found to show strong anti-hepatocarcinoma activities. However, the anticancer mechanism remains not clear. To search for the potential biomarkers for the evaluation of treatment, ^{1}H NMR metabolomics was employed to distinguish the serum metabolic profiling of the RPS treatment group from that of the model group [52]. As a result, RPS decreased the serum levels of lactate, acetate, N-acetyl amino acid, and glutamine, which has shown that RPS was a potential anticancer drug by inhibiting the aerobic glycolysis, lipogenesis, and glutamine metabolism. As one of the rarest plants, *Camellia nitidissima Chi* was reported to have various pharmacological activities, including anti-CRC. However, its anti-CRC efficacies remained to be confirmed due to its complex components and underlying complicated mechanisms. To address these issues, Li and his colleagues employed ^{1}H NMR-based metabolomics of the intestine, kidney, and spleen from azoxymethane/dextran sodium sulfate (AOM/DSS)-induced CRC mice model [53]. They found that *C. nitidissima Chi* extracts could markedly suppress AOM/DSS-induced CRC via reversing the disturbed metabolic profiling to the normal state. What is more, compared with the water-soluble fraction of *C. nitidissima Chi*, its butanol fraction exhibited a better efficacy. Gnb was widely used in the treatment of lung carcinoma (LLC) with increasing drug resistance and serious metabolic disorders. Si Jun Zi Tang (SJZ) is a four-herb Chinese medicine formula and has shown potential of anticancer properties. To explore the underlying mechanisms of the co-therapy of Gnb and SJZ, Li et al. conducted an integrated network pharmacology and Q-TOF LC/MS-based metabolomics of plasma from LLC-bearing mice model in vivo [54]. SJZ was shown to increase the anti-LLC effects of Gnb via restoring TCA cycle, linoleic acid metabolism, and tyrosine and tryptophan metabolism, revealing that co-therapy of Gnb and SJZ may increase the anti-LLC potential of Gnb.

2.2 Lipid metabolism

Besides the glucose and energy metabolisms having an essential role in the tumorigenesis process of cancer, it has been also reported that lipid metabolism such as de novo lipogenesis regulates the synthesis of cellular membranes and the important signaling molecules of rapidly proliferating tumor cells [55]. Targeting the lipid metabolism could be a novel therapeutics for cancer treatment. Here the recent metabolomics studies of pure compounds, herb extracts, and formulations from Chinese medicines, which target lipid metabolism, have been reviewed.

Flexibilide is a natural compound derived from the soft coral *Sinularia flexibilis* with tumor inhibitory effects. To clarify the pharmacological mechanism, a UPLC/Q-TOF MS-based metabolomics of HCT-116 cells in vitro on colon cancer was conducted [56]. It was found that flexibilide treatment greatly elevated lysophosphatidylcholine (LysoPC) and diminished phosphocholine and phosphatidylcholine (PC), revealing that flexibilide exhibited the therapeutic effect on colon cancer mainly via downregulating PC biosynthesis pathway. Englerin A is a guaiane sesquiterpene derived from the plant *Phyllanthus engleri* with potential antineoplastic property. To uncover the therapeutic role of englerin A on clear cell renal carcinoma, Batova et al. conducted a LC-MS/MS-based metabolomics of A498 cells in vitro [57]. It was found that englerin A significantly reversed lipid metabolism and increased ceramide levels. Then the increasing ceramides inhibited renal carcinoma cells. Isoquercitrin is a kind of flavonoid derived from various plants, such as *Psidium guajava* and *Fagopyrum tataricum*. It has potential antitumor activities. To decipher its therapeutic role in bladder cancer, a UPLC/Q-TOF MS-based metabolomics of T24 cells in vitro was conducted [58]. Isoquercitrin treatment was found to regulate lipid and anaerobic glycolysis via activating the AMPK pathway. Peiminine is an active substance derived from the bulbs of *Fritillaria thunbergii* with potential antineoplastic property against CRC. To investigate the molecular mechanisms of how it worked, a combined UPLC-MS- and GC/MS-based metabolomics of HCT-116 cells in vitro was used [59]. Peiminine treatment altered several metabolites, including lignocerate (24:0), oleate (18:1n9), glutamine, and glucose, indicating peiminine exerted the predominant therapeutic effect mainly via the metabolic regulation of lipids, amino acids, and carbohydrates. 8u is an acridine derivative with potential antiproliferative activity against cancer. To explore its therapeutic effects on HCC, a combined proteomics and UPLC/Q-TOF MS-based metabolomics of HepG2 cells in vitro was used [60]. 8u was found to significantly inhibit the invasion and metastasis of HepG2 cells and regulate intracellular lipid metabolism mainly via suppressing the PI3K/Akt signaling pathway. Genistein is a kind of isoflavone with antineoplastic property. However, high concentration of genistein shows promotional role in cancer. Calcitriol (1α,25(OH)2 vitamin D3) is a primary bioactive hormonal form of vitamin D3. It also shows the antitumor effect. To explore the synergism effects of co-therapy of genistein and calcitriol on osteosarcoma, a GC/MS-based metabolomics of MG-63 cells in vitro was conducted [61]. Co-therapy of genistein and calcitriol was found to regulate lipids and amino acids rather than energy metabolism. Taken together, the promotional effects of high level of genistein on osteosarcoma could be decreased by the co-treatment of calcitriol. Silibinin, as a kind of natural flavonoid, is derived from the milk thistle (*Silybum marianum*) seeds with strong hepatoprotective activity. To clarify the pharmacological mechanism of how silibinin exerted antineoplastic property, a ^{1}H-NMR-based metabolomics of tumor tissues from 22Rv1 xenograft model in vivo was used [62]. Silibinin treatment did not greatly affect glucose uptake of PCa tumor but decreased the lipid synthesis via suppressing hypoxia-induced signaling. Acyclic retinoid (ACR), as a synthetic vitamin A-like compound, exhibits antineoplastic property against HCC. To decipher the molecular mechanisms, comprehensive cationic and lipophilic metabolomics of liver tissues from mouse DEN-induced HCC model in vivo was conducted by CE-TOFMS and LC-TOFMS [63]. ACR predominantly reversed lipogenesis but not glucose metabolism by inhibiting linoleic acid metabolites, revealing lipid metabolic reprogramming played a critical role in the protective effects of ACR on HCC.

Soft coral, *Sinularia* sp., is reported to show potential antineoplastic property. To decipher the molecular mechanisms, a MS-based metabolomics of Hep 3B cells in vitro was conducted [64]. It was found that the *Bornean Sinularia* sp. extract could regulate the sphingolipids and ceramide, revealing that the regulation of dysregulated lipids may account for the antineoplastic property of *Bornean Sinularia* sp. against HCC. *Forsythiae Fructus* (FAE), as the dry fruit of *Forsythia suspensa* (*Thunb.*) Vahl. of Oleaceae family, shows potential anticancer properties. To characterize in detail the action mechanism, Bao et al. conducted a UPLC/Q-TOF MS-based metabolomics of serum from B16-F10 melanoma-bearing mice model in vivo [65]. Aqueous extract of FAE was found to restore the disturbed metabolic profile by increasing several LysoPCs in glycerophospholipid metabolisms, revealing that the regulation of glycerophospholipid metabolisms may have an essential role in the antineoplastic property of FAE. Nutmeg is a seed of the fruit of *Myristica fragrans* with antimicrobial and anticancer activities. To explore the role of its antimicrobial activity in cancer protection, a UPLC/ESI-QTOF-MS-based metabolomics of serum from colon cancer model was investigated [66]. Nutmeg extract treatment was found to regulate lipid metabolism by decreasing four uremic toxins generated from the gut microbiota, revealing that the regulation of lipid metabolism and gut microbiota may be an effective therapy for colon cancer treatment. Volatile oil is extracted from *Saussurea lappa Decne* (VOSL), and costunolide and dehydrocostus lactone (Cos–Dehy), accounting for almost 75% of VOSL by weight, are the primary active chemical compositions of VOSL. It has been reported that they all can suppress the MCF-7cells in vitro. To characterize in detail the action mechanism of how they worked, a combined GC × GC–TOF/MS and UPLC/Q-TOF MS metabolomics of serum and urine from MCF-7 xenograft mice in vivo was conducted [67]. It was revealed that both VOSL and Cos–Dehy could relieve metabolic disturbance by decreasing glycolysis and steroid hormone metabolism and increasing unsaturated fatty acids metabolism, suggesting that VOSL is a potential therapeutics against breast cancer. Shuihonghuazi formula (SHHZF) is a famous formula which has been widely used clinically for the treatment of liver cancer. To explore its action mechanism, a DEN-induced HCC rat model was built, and a HPLC/ESI-TOF-MS-based metabolomics of plasma from this model was conducted [68]. SHHZF was found to elevate the levels of arachidonic acid-like substances and the shift of phosphatidylethanolamine (PE) to PC, revealing the reversion of the disturbed fatty acid and bile acid metabolism played an important role in the therapeutic effects of SHHZF on HCC. Qi-Yu-San-Long Decoction (QYSLD) is a classic formula, which has been widely used clinically for LLC treatment. To characterize in detail the action mechanism of how it works, a UPLC/Q-TOF MS-based metabolomics was conducted [69]. Lewis LLC mice model was firstly built, and plasma was collected for metabolomics analysis. QYSLD was found to regulate sphingolipid metabolism, glycerophospholipid metabolism, arachidonic acid metabolism, fatty acid degradation, and steroid hormone biosynthesis. *Rhizoma Curcumae* and *Rhizoma Sparganii* (RCRS) is a famous Chinese medicine drug pair to treat hysteromyoma. To investigate the molecular mechanisms of how this drug pair works on hysteromyoma, a UPLC/Q-TOF MS-based metabolomics was conducted using the serum and urine from hysteromyoma rat model [70]. RCRS treatment characterized 16 and 18 potential biomarkers from serum and urine, respectively, which were associated with glyoxylate, dicarboxylate, and linoleic acid metabolisms.

2.3 Amino acid metabolism

As mentioned above, besides the consumption of glucose, cancer cells have also been reported to favor glutamine as a preferential fuel. Glutamine metabolism has an essential role in the pathological progression of cancer and could be a potential therapeutic option for cancer. Besides the key metabolite glutamine, it has been reported many other amino acids also play an essential role in cancer.

Delta-tocotrienol (δT) is one of the isomers of vitamin E with antineoplastic property. To explore underlying action mechanism, a ^{1}H-NMR-based metabolomics of A549 and H1299 cells in vitro was used [71]. In detail, δT treatment could suppress the glutamine uptake via suppressing glutamine transporters and then resulted in the induction of apoptosis and suppression of cell proliferation. Celastrol is a bioactive compound derived from *Trypterygium wilfordii HOOK F.* with potential antineoplastic property. To explore underlying action mechanism involved in its anti-colon cancer activity, a UPLC/MS-based metabolomics of HCT116 cells in vitro was conducted [72]. Metabolomics analysis found celastrol changed the levels of lipid markers, carnitine, and amino acids. Tryptophan was further identified as special biomarker by targeted metabolite analysis. Melittin, as a cytotoxic peptide isolated from bee venom, was shown to sensitize the response of ovarian cancer cells to cisplatin treatment. To explore an underlying action mechanism, a LC-MS metabolomics of A2780 and A2780CR cell lines in vitro was employed [73]. It was found that melittin treatment of cisplatin-sensitive cells decreased glutamine, proline, and arginine pathways. Chlorogenic acid (ChA) and caffeic acid (CaA), both as a kind of polyphenol, have shown anti-HCC activities. To decipher the molecular mechanisms, a combined 16S rRNA and metabolomics was conducted [74]. It was found that both CaA and ChA treatments reverse 28 metabolites. In detail, the levels of ethanolamine, L-methionine, L-tyrosine, and bilirubin were associated with diminished *Prevotella* 9 and *Lachnospiraceae incertae sedis* and elevated *Rumincoccaceae* UCG-004. Lodi et al. used untargeted metabolomics and metabolic flux analysis to investigate the synergistic effects of resveratrol, curcumin, and ursolic acid [75]. It was found that compared with the individual treatment, the combined treatment had the greater antineoplastic property. Mechanically, glutamine metabolism was regulated by the compound combinations.

Polyphenols are characterized as a hydroalcoholic chestnut shell extract. Sorice et al. used ^{1}H-NMR-based metabolomics of HepG2 cells in vitro to study the anti-HCC activity of polyphenols extracted from chestnut shell (PECS) [76]. PECS was found to regulate the levels of some amino acids. Annonaceous acetogenins (ACGs) are a group of C35 or C37 secondary metabolites isolated from plants in *Annonaceae*. To explore underlying action mechanism of the anti-HCC activity of ACGs, a UPLC-ESI-Q-TOF-MS-based metabolomics of SMMC 7721 cells in vitro was conducted [77]. ACG treatment could regulate the metabolisms of sphingolipid, arginine, glutathione, and proline, which further reversed the resistance of

SMMC 7721 cells to adriamycin. *Hedyotis diffusa* is a famous Chinese herbal medicine with antineoplastic property. To predict the potential underlying mechanism, a ^{1}H NMR-based metabolomics was conducted to use plasma and urine from rats bearing Walker 256 tumor [78]. *Hedyotis diffusa* treatment was found to reverse lactate, acetate, choline, 3-hydroxybutyrate, and L-glutamine in plasma as well as creatinine, L-aspartate, N-acetyl-L-aspartate, and ornithine in urine. Wang et al. developed a combined gut microbiota and metabolomics analysis to investigate the anti-CRC activity of *American ginseng* [79]. By GC/TOF-MS-based metabolomics, *American ginseng* was found to regulate the metabolisms of carbohydrates, lipids, and amino acids. By the 16S rRNA data analysis, *American ginseng* was found to inhibit the changes of microbiome community caused by azoxymethane/dextran sulfate sodium. Kushen injection (CKI) is a famous Chinese medicine preparation and widely used for treating multiple kinds of solid tumors. To evaluate the anti-HCC mechanisms of CKI, a combined network analysis and ^{1}H-NMR-based metabolomics were used [80]. Network pharmacology analysis found the primary active compounds, the potential targets, and pathways associated with the anti-HCC effects of CKI, which was further validated by metabolomics. Metabolomics analysis validated the primary pathways associated with the anti-HCC effects of CKI were amino acid metabolism and glycometabolism.

2.4 Nucleotide metabolism

To support the rapid proliferation of cancer cells, nucleic acid synthesis is shown to be accelerated. Accordingly, the anticancer therapy targeting nucleotide metabolism has obtained numerous attentions. Here the recent metabolomics studies of Chinese medicines targeting nucleotide metabolism have been reviewed.

Glaucocalyxin A (GLA) is an ent-kaurene diterpenoid derived from *Rabdosia japonica* and has shown to have antineoplastic property. To explore underlying action mechanism underlying the anti-HCC activity of GLA, a combined GC/MS- and LC/MS-based metabolomics was conducted using SMMC7721 cells in vitro [81]. It was found GLA treatment diminished amino acid metabolism and elevated the metabolisms of sphingolipid, purine, and pyrimidine. Taurine, as the most abundant free amino acid, has the antineoplastic property against breast cancer. To elucidate the mechanisms underlying the therapeutic benefits of taurine against breast cancer, a GC–TOF-MS-based metabolomics of plasma from dimethylbenz[a]anthracene-induced breast carcinogenesis in rats was conducted [82]. It was found that taurine treatment regulated 23 differential metabolites, which were associated with glucose, energy and amino acid, as well as nucleic acid metabolisms. Celastrol is a bioactive compound derived from *Trypterygium wilfordii HOOK F.* with potential antineoplastic property. To explore underlying action mechanism involved in its anti-acute promyelocytic leukemia activity, a UPLC-MS-based metabolomics of HL-60 cells in vitro and tumor tissue from mice xenograft model in vivo was conducted [83]. It was found that celastrol treatment regulated uridine metabolite, which further enhanced apoptosis. The development of radioprotector to reduce the serious side effects and complications caused by radiotherapy is important. Gamma-tocotrienol (GT3) is one of the isomers of vitamin E with antineoplastic property. To explore the radioprotective mechanism of GT3, a UPLC-QTF MS-based metabolomics of serum from nonhuman primate models in vivo was conducted [84]. It was found that GT3 could regulate the changed fatty acid beta-oxidation and amino acid and purine catabolism metabolisms caused by irradiation.

Red kidney bean, also named as *Phaseolus vulgaris L.*, possesses antineoplastic property. To evaluate its anti-melanoma activity, a combined network pharmacology and LC-MS-based metabolomics analysis was conducted using B16-F10 cells in vitro [85]. It was found that the kernel of red kidney bean (RKBC) extract treatment markedly elevated cellular level of cGMP. Network pharmacology analysis showed that quercetin might act as the main effective ingredient of RKBC extract. Ku-jin tea (KJT) is a famous beverage derived from the leaves of the plant *Acer tataricum subsp. ginnala* with antineoplastic property. A UPLC/Q-TOF MS-based metabolomics of urine from azoxymethane-induced precancerous colorectal lesion model in rats was conducted to investigate molecular modes of inhibitory effects of KJT against CRC [86]. It was found that KJT treatment modulated amino acid and purine metabolisms, which accounted for the chemopreventive effects of KJT.

2.5 Other related metabolisms

Except for the anticancer therapies of Chinese medicine targeting the changed metabolisms mentioned above, there are also some other related metabolisms which are the targets by Chinese medicine. Fisetin is a kind of plant flavonoid with antineoplastic property. A HPLC/ESI-MS-based metabolomics of tumor tissues from PCa xenografts in vivo was conducted to explore its therapeutic benefit for PCa [87]. Fisetin treatment was shown to downregulate secreted and intracellular hyaluronan (HA), which conferred resistance to prostate oncogenesis. Yang et al. developed a LC-MS/MS-based metabolomics to study the bioefficacy of a plant galactolipid 1,2-di-O-α-linolenoyl-3-O-β-D- galactopyranosyl-sn-glycerol (dLGG) against melanoma [88]. dLGG treatment markedly elevated 12/15-LOX catalyzed oxylipin products in serum, revealing the novel therapeutic mechanism of phytoagent dLGG against melanoma. Derived from the medicinal plant *Elephantopus scaber*, deoxyelephantopin (DET) is a germacranolide sesquiterpene lactone with antineoplastic property. To study whether the co-therapy of DET and cisplatin could reduce the cisplatin-induced nephrotoxicity, a UPLC/ESI-QTOF MS-based metabolomics of kidney tissues from murine B16 metastatic allograft model in vivo was conducted [89]. It was shown that co-therapy of DET and cisplatin could reverse the changed urea cycle metabolites and hippuric acid in renal tissues caused by cisplatin, revealing that the co-therapy of DET and cisplatin could be an effective treatment with low toxicity for melanoma.

Liu et al. developed a UHPLC-MS/MS-based targeted metabolomics to evaluate the efficacy of anticancer drugs, including a traditional Chinese medicine injection Aidi injections and fluorouracil [90]. It was found that with the progression of squamous cell carcinoma of the lung, the levels of 1,3-diaminopropane, cadaverine, and N-acetylputrescine altered. The two-drug treatment alone or co-therapy reversed the levels of 1,3-diaminopropane, cadaverine, and N-acetylputrescine. The team also used this metabolomics method to evaluate the efficacy of Aidi injections on CRC [91]. It was found that Aidi injection treatment could reverse polyamine metabolism, especially agmatine and putrescine, revealing that plasma polyamine could be a biomarker for both early diagnosis and medical treatment of CRC.

3. Current perspectives and future challenges

In accordance with the holistic perspective of Chinese medicines, metabolomics can help to explain the underlying mechanisms of the anticancer effects of Chinese medicines or the reversion of the drug resistance of chemotherapy and radiotherapy. It can also help to rapidly compare the anticancer effects of multiple compounds from Chinese medicines and act as a quick preliminary platform to screen the most dominant compound related to anticancer bioactivity. Based on the metabolomics analyses of modern studies of Chinese medicines with antineoplastic properties, the potential of metabolic treatment with pure compounds, herb extracts, and formulations from Chinese medicines is gaining numerous attentions. However, many challenges still exist in the metabolomics study of antineoplastic Chinese medicines, and there is still a long way for the wide application of metabolomics of Chinese medicines into the treatment of cancer. Firstly, it is critical to make good experimental design before starting the experiment, such as the choices of samples, analytical platforms, and methodological approaches. Secondly, it is quite essential for researchers to develop metabolomics, such as the development of data excavation and the identification and quantification of more metabolites. Thirdly, it is important for us to validate the results from metabolomics studies with more mechanical biological experiments. Fourthly, as no one single technology could achieve a comprehensive result, it is strongly suggested to combine metabolomics with some other advanced technologies for better investigation of the action mechanisms of antineoplastic Chinese medicines, such as other "omics" technologies, network pharmacology, and gut microbiome analyses. Last but not least, more attentions will be drawn to personalized treatment based on metabolomics. It has been reported that because of the interaction between genes and environment (polypharmacy, gut microbiota, xenobiotics), not all patients present the same response to drug treatment [92]. Personalized treatment has been put forward and of great importance nowadays. Although pharmacogenomics is still the only means in terms of personalized treatment, its limitation of ignoring the environmental influences has been increasingly recognized. As an alternative and complementary manner, pharmacometabolomics is an emerging "omics" and has been proposed for personalized treatment [16]. As the results of both genetic and environmental influences, pharmacometabolomics can help to understand individual phenotypic variations in drug responses by providing individual metabolic signatures of both gene-derived endogenous and environment-derived exogenous metabolites [93]. Pharmacometabolomics will offer an intriguingly avenue for personalized treatment in the future.

4. Conclusions

In this chapter, we systematically reviewed recent studies on metabolism-targeting anticancer therapies based on metabolomics in terms of glucose, lipid, amino acid, and nucleotide metabolisms and other altered metabolisms, with special emphasis on the potential of metabolic treatment with pure compounds, herb extracts, and formulations from Chinese medicines. The trends of future development of metabolism-targeting anticancer therapies were also discussed. Hopefully, we expect that through the systematic review on the recent metabolomics studies targeting Chinese medicine treatment on human cancers, more attention will be drawn to the promising candidates from the resourceful Chinese medicine as effective neoadjuvant therapies for cancer treatment clinically.

Acknowledgements

The study was financially supported by grants from the research council of the University of Hong Kong (Project Codes: 104004092, 104004460, 104004746), the Research Grants Committee (RGC) of Hong Kong, HKSAR (Project Codes: 764708, 766211, 17152116), Wong's Donation on Modern Oncology of Chinese Medicine (Project code: 200006276), Gala Family Trust (Project Code: 200007008), Innovation Technology Fund of Hong Kong (ITF. Project code: 260900263), and HMRF (Project code: 16172751).

Author details

Wei Guo, Hor-Yue Tan, Ning Wang and Yibin Feng*
School of Chinese Medicine, Li Ka Shing Faculty of Medicine, The University of Hong Kong, Hong Kong, China

*Address all correspondence to: yfeng@hku.hk

References

[1] Wang Z et al. An update on Chinese herbal medicines as adjuvant treatment of anticancer therapeutics. Bioscience Trends. 2018;**12**(3):220-239

[2] Chen W et al. National cancer incidence and mortality in China, 2012. Chinese Journal of Cancer Research. 2016;**28**(1):1-11

[3] Qi F et al. The advantages of using traditional Chinese medicine as an adjunctive therapy in the whole course of cancer treatment instead of only terminal stage of cancer. Bioscience Trends. 2015;**9**(1):16-34

[4] Lou JS, Yao P, Tsim KWK. Cancer treatment by using traditional chinese medicine: Probing active compounds in anti-multidrug resistance during drug therapy. Current Medicinal Chemistry. 2018;**25**(38):5128-5141

[5] Ward PS, Thompson CB. Metabolic reprogramming: A cancer hallmark even warburg did not anticipate. Cancer Cell. 2012;**21**(3):297-308

[6] Warburg O, Wind F, Negelein E. The metabolism of tumors in the body. The Journal of General Physiology. 1927;**8**(6):519-530

[7] Son J et al. Glutamine supports pancreatic cancer growth through a KRAS-regulated metabolic pathway. Nature. 2013;**496**(7443):101-105

[8] Selak MA et al. Succinate links TCA cycle dysfunction to oncogenesis by inhibiting HIF-alpha prolyl hydroxylase. Cancer Cell. 2005;**7**(1):77-85

[9] Tong WH et al. The glycolytic shift in fumarate-hydratase-deficient kidney cancer lowers AMPK levels, increases anabolic propensities and lowers cellular iron levels. Cancer Cell. 2011;**20**(3):315-327

[10] Dang L, Yen K, Attar EC. IDH mutations in cancer and progress toward development of targeted therapeutics. Annals of Oncology. 2016;**27**(4):599-608

[11] Lee M, Ko H, Yun M. Cancer metabolism as a mechanism of treatment resistance and potential therapeutic target in hepatocellular carcinoma. Yonsei Medical Journal. 2018;**59**(10):1143-1149

[12] Vernieri C et al. Targeting cancer metabolism: Dietary and pharmacologic interventions. Cancer Discovery. 2016;**6**(12):1315-1333

[13] Jiang CY et al. A (1)H NMR-based metabonomic investigation of time-related metabolic trajectories of the plasma, urine and liver extracts of hyperlipidemic hamsters. PLoS One. 2013;**8**(6):e66786

[14] Guo W et al. Quantitative metabolomic profiling of plasma, urine, and liver extracts by (1)H NMR spectroscopy characterizes different stages of atherosclerosis in hamsters. Journal of Proteome Research. 2016;**15**(10):3500-3510

[15] Ellis DI et al. Metabolic fingerprinting as a diagnostic tool. Pharmacogenomics. 2007;**8**(9):1243-1266

[16] Wishart DS. Emerging applications of metabolomics in drug discovery and precision medicine. Nature Reviews Drug Discovery. 2016;**15**(7):473-484

[17] Zhang A et al. Modern analytical techniques in metabolomics analysis. Analyst. 2012;**137**(2):293-300

[18] Dunn WB, Bailey NJ, Johnson HE. Measuring the metabolome: current analytical

technologies. Analyst. 2005;**130**(5):606-625

[19] Trezzi JP, Vlassis N, Hiller K. The role of metabolomics in the study of cancer biomarkers and in the development of diagnostic tools. Advances in Experimental Medicine and Biology. 2015;**867**:41-57

[20] Fan TW et al. Stable isotope-resolved metabolomics and applications for drug development. Pharmacology & Therapeutics. 2012;**133**(3):366-391

[21] Spratlin JL, Serkova NJ, Eckhardt SG. Clinical applications of metabolomics in oncology: A review. Clinical Cancer Research. 2009;**15**(2):431-440

[22] Griffin JL, Shockcor JP. Metabolic profiles of cancer cells. Nature Reviews Cancer. 2004;**4**(7):551-561

[23] Guo W et al. Deciphering hepatocellular carcinoma through metabolomics: From biomarker discovery to therapy evaluation. Cancer Management and Research. 2018;**10**:715-734

[24] Wang N et al. Berberine and Coptidis Rhizoma as potential anticancer agents: Recent updates and future perspectives. Journal of Ethnopharmacology. 2015;**176**:35-48

[25] Wang X, Fang G, Pang Y. Chinese medicines in the treatment of prostate cancer: From formulas to extracts and compounds. Nutrients. 2018;**10**(3):283

[26] Wang N et al. F-actin reorganization and inactivation of rho signaling pathway involved in the inhibitory effect of Coptidis Rhizoma on hepatoma cell migration. Integrative Cancer Therapies. 2010;**9**(4):354-364

[27] Wang N et al. Fangchinoline induces autophagic cell death via p53/sestrin2/AMPK signalling in human hepatocellular carcinoma cells. British Journal of Pharmacology. 2011;**164**(2b):731-742

[28] Wang N et al. Berberine induces autophagic cell death and mitochondrial apoptosis in liver cancer cells: The cellular mechanism. Journal of Cellular Biochemistry. 2010;**111**(6):1426-1436

[29] Wang N et al. Up-regulation of TIMP-1 by genipin inhibits MMP-2 activities and suppresses the metastatic potential of human hepatocellular carcinoma. PLoS One. 2012;**7**(9):e46318

[30] Tan HY et al. Autophagy-induced RelB/p52 activation mediates tumour-associated macrophage repolarisation and suppression of hepatocellular carcinoma by natural compound baicalin. Cell Death & Disease. 2015;**6**:e1942

[31] Ma X et al. Discovery of traditional Chinese medicine monomers and their synthetic intermediates, analogs or derivatives for battling P-gp-mediated multi-drug resistance. European Journal of Medicinal Chemistry. 2018;**159**:381-392

[32] Zhao HD et al. Research progress on reversing multidrug resistance in tumors by using Chinese medicine. Chinese Journal of Integrative Medicine. 2018;**24**(6):474-480

[33] Armitage EG, Ciborowski M. *Applications of metabolomics in cancer studies*. Advances in Experimental Medicine and Biology. 2017;**965**:209-234

[34] Pandey R et al. Metabolomic signature of brain cancer. Molecular Carcinogenesis. 2017;**56**(11):2355-2371

[35] McCartney A et al. Metabolomics in breast cancer: A decade in review. Cancer Treatment Reviews. 2018;**67**:88-96

[36] Iwao C, Shidoji Y. Upregulation of energy metabolism-related, p53-target TIGAR and SCO2 in HuH-7 cells with p53 mutation by geranylgeranoic acid treatment. Biomedical Research. 2015;**36**(6):371-381

[37] Chen GQ et al. Halofuginone inhibits colorectal cancer growth through suppression of Akt/mTORC1 signaling and glucose metabolism. Oncotarget. 2015;**6**(27):24148-24162

[38] Gao L et al. (1)H nuclear magnetic resonance based metabolomics approach reveals the metabolic mechanism of (-)-5-Hydroxy-equol against hepatocellular carcinoma cells in vitro. Journal of Proteome Research. 2018;**17**(5):1833-1843

[39] Younis T et al. Nummularic acid, a triterpenoid, from the medicinal plant Fraxinus xanthoxyloides, induces energy crisis to suppress growth of prostate cancer cells. Molecular Carcinogenesis. 2018;**57**(10):1267-1277

[40] Sun H et al. Cell metabolomics identify regulatory pathways and targets of magnoline against prostate cancer. Journal of Chromatography B. 2018;**1102-1103**:143-151

[41] Abu El Maaty MA et al. 1,25(OH)2D3 disrupts glucose metabolism in prostate cancer cells leading to a truncation of the TCA cycle and inhibition of TXNIP expression. Biochimica et Biophysica Acta, Molecular Cell Research. 2017;**1864**(10):1618-1630

[42] Yun J et al. Vitamin C selectively kills KRAS and BRAF mutant colorectal cancer cells by targeting GAPDH. Science. 2015;**350**(6266):1391-1396

[43] Silvers MA et al. The NQO1 bioactivatable drug, beta-lapachone, alters the redox state of NQO1+ pancreatic cancer cells, causing perturbation in central carbon metabolism. The Journal of Biological Chemistry. 2017;**292**(44):18203-18216

[44] Arminan A et al. HIF-1alpha inhibition by diethylstilbestrol and its polyacetal conjugate in hypoxic prostate tumour cells: insights from NMR metabolomics. Journal of Drug Targeting. 2017;**25**(9-10):845-855

[45] Liberti MV et al. A predictive model for selective targeting of the Warburg effect through GAPDH inhibition with a natural product. Cell Metabolism. 2017;**26**(4):648-659. e8

[46] Zhu S et al. Metabolic shift induced by omega -3 PUFAs and rapamycin lead to cancer cell death. Cellular Physiology and Biochemistry. 2018;**48**(6):2318-2336

[47] Qiu P et al. Curcumin attenuates N-nitrosodiethylamine-induced liver injury in mice by utilizing the method of metabonomics. Journal of Agricultural and Food Chemistry. 2017;**65**(9):2000-2007

[48] Mishra P et al. 6,7-dimethoxy-1,2,3,4-tetrahydro-isoquinoline-3-carboxylic acid attenuates colon carcinogenesis via blockade of IL-6 mediated signals. Biomedicine & Pharmacotherapy. 2018;**100**:282-295

[49] Ma T et al. Metabonomics applied in exploring the antitumour mechanism of physapubenolide on hepatocellular carcinoma cells by targeting glycolysis through the Akt-p53 pathway. Scientific Reports. 2016;**6**:29926

[50] Parashar P et al. Biotinylated naringenin intensified anticancer effect of gefitinib in urethane-induced lung cancer in rats: favourable modulation of apoptotic regulators and serum metabolomics. Artificial Cells, Nanomedicine, and Biotechnology. 2018;**46**(sup3):S598-S610

[51] Lovelace ES et al. Silymarin suppresses cellular inflammation by inducing reparative stress signaling. Journal of Natural Products. 2015;**78**(8):1990-2000

[52] Qiu P et al. Utilization of metabonomics to identify serum biomarkers in murine H22 hepatocarcinoma and deduce antitumor mechanism of Rhizoma Paridis saponins. Chemico-Biological Interactions. 2016;**256**:55-63

[53] Li MH et al. Nuclear magnetic resonance-based metabolomics approach to evaluate the prevention effect of Camellia nitidissima Chi on Colitis-associated carcinogenesis. Frontiers in Pharmacology. 2017;**8**:447

[54] Li C et al. The modulatory properties of Si Jun Zi Tang enhancing anticancer of gefitinib by an integrating approach. Biomedicine & Pharmacotherapy. 2019;**111**:1132-1140

[55] Currie E et al. Cellular fatty acid metabolism and cancer. Cell Metabolism. 2013;**18**(2):153-161

[56] Gao D et al. Metabolomics study on the antitumor effect of marine natural compound flexibilide in HCT-116 colon cancer cell line. Journal of Chromatography B. 2016;**1014**:17-23

[57] Batova A et al. Englerin A induces an acute inflammatory response and reveals lipid metabolism and ER stress as targetable vulnerabilities in renal cell carcinoma. PLoS One. 2017;**12**(3):e0172632

[58] Wu P et al. Apoptosis triggered by isoquercitrin in bladder cancer cells by activating the AMPK-activated protein kinase pathway. Food & Function. 2017;**8**(10):3707-3722

[59] Zheng Z et al. Peiminine inhibits colorectal cancer cell proliferation by inducing apoptosis and autophagy and modulating key metabolic pathways. Oncotarget. 2017;**8**(29):47619-47631

[60] Wang N et al. 8u, a pro-apoptosis/cell cycle arrest compound, suppresses invasion and metastasis through HSP90alpha downregulating and PI3K/Akt inactivation in hepatocellular carcinoma cells. Scientific Reports. 2018;**8**(1):309

[61] Engel N et al. Synergistic action of Genistein and Calcitriol in immature osteosarcoma MG-63 cells by SGPL1 up-regulation. PLoS One. 2017;**12**(1):e0169742

[62] Deep G et al. Silibinin inhibits hypoxia-induced HIF-1alpha-mediated signaling, angiogenesis and lipogenesis in prostate cancer cells: In vitro evidence and in vivo functional imaging and metabolomics. Molecular Carcinogenesis. 2017;**56**(3):833-848

[63] Qin XY et al. Metabolome analyses uncovered a novel inhibitory effect of acyclic retinoid on aberrant lipogenesis in a mouse diethylnitrosamine-induced hepatic tumorigenesis model. Cancer Prevention Research. 2016;**9**(3):205-214

[64] Ling YS et al. MS-based metabolomics revealing Bornean Sinularia sp. extract dysregulated lipids triggering programmed cell death in Hepatocellular carcinoma. Natural Product Research. Dec 2018;**26**:1-8

[65] Bao J et al. Anti-melanoma activity of Forsythiae Fructus aqueous extract in mice involves regulation of glycerophospholipid metabolisms by UPLC/Q-TOF MS-based metabolomics study. Scientific Reports. 2016;**6**:39415

[66] Li F et al. Modulation of colon cancer by nutmeg. Journal of Proteome Research. 2015;**14**(4):1937-1946

[67] Peng ZX et al. Metabolic transformation of breast cancer in a

MCF-7 xenograft mouse model and inhibitory effect of volatile oil from Saussurea lappa Decne treatment. Metabolomics. 2015;**11**(3):636-656

[68] Bao Y et al. Metabolomic study of the intervention effects of Shuihonghuazi formula, a traditional Chinese medicinal formulae, on hepatocellular carcinoma (HCC) rats using performance HPLC/ESI-TOF-MS. Journal of Ethnopharmacology. 2017;**198**:468-478

[69] Wu H et al. Intervention effect of Qi-Yu-San-Long Decoction on Lewis lung carcinoma in C57BL/6 mice: Insights from UPLC-QTOF/MS-based metabolic profiling. Journal of Chromatography B. 2018;**1102-1103**:23-33

[70] Li W et al. An integrated serum and urinary metabonomic research of Rhizoma Curcumae-Rhizoma Sparganii drug pair in hysteromyoma rats based on UPLC-Q-TOF-MS analysis. Journal of Ethnopharmacology. 2019;**231**:374-385

[71] Rajasinghe LD, Hutchings M, Gupta SV. Delta-tocotrienol modulates glutamine dependence by inhibiting ASCT2 and LAT1 transporters in non-small cell lung cancer (NSCLC) cells: A metabolomic approach. Metabolites. 2019;**9**(3):50

[72] Qi Y et al. Celastrol suppresses tryptophan catabolism in human colon cancer cells as revealed by metabolic profiling and targeted metabolite analysis. Biological & Pharmaceutical Bulletin. 2018;**41**(8):1243-1250

[73] Alonezi S et al. Metabolomic profiling of the effects of melittin on cisplatin resistant and cisplatin sensitive ovarian cancer cells using mass spectrometry and biolog microarray technology. Metabolites. 2016;**6**(4):35

[74] Zhang Z et al. Metabolic and microbial signatures in rat hepatocellular carcinoma treated with caffeic acid and chlorogenic acid. Scientific Reports. 2017;**7**(1):4508

[75] Lodi A et al. Combinatorial treatment with natural compounds in prostate cancer inhibits prostate tumor growth and leads to key modulations of cancer cell metabolism. NPJ Precision Oncology. 2017;**1**:18

[76] Sorice A et al. Potential anticancer effects of polyphenols from chestnut shell extracts: Modulation of cell growth, and cytokinomic and metabolomic profiles. Molecules. 2016;**21**(10):1411

[77] Ma C et al. Non-targeted metabolomic analysis on multidrug resistance hepatocellular carcinoma cell and reversal effect of annonaceous acetogenins. Journal of Pharmaceutical and Biomedical Analysis. 2019;**164**:489-495

[78] Wang Z et al. Metabolic effects of Hedyotis diffusa on rats bearing Walker 256 tumor revealed by NMR-based metabolomics. Magnetic Resonance in Chemistry. 2018;**56**(1):5-17

[79] Wang CZ et al. American ginseng attenuates colitis-associated colon carcinogenesis in mice: Impact on gut microbiota and metabolomics. Cancer Prevention Research. 2016;**9**(10):803-811

[80] Gao L et al. Uncovering the anticancer mechanism of compound Kushen injection against HCC by integrating quantitative analysis, network analysis and experimental validation. Scientific Reports. 2018;**8**(1):624

[81] Liu Y et al. Effects of glaucocalyxin A on human liver cancer cells as revealed by GC/MS- and LC/MS-based metabolic profiling. Analytical and Bioanalytical Chemistry. 2018;**410**(14):3325-3335

[82] He YU, Li QQ, Guo SC. Taurine attenuates dimethylbenz[a]anthracene-induced breast tumorigenesis in rats: A plasma metabolomic study. Anticancer Research. 2016;**36**(2):533-543

[83] Zhang X et al. Metabolomics profiles delineate uridine deficiency contributes to mitochondria-mediated apoptosis induced by celastrol in human acute promyelocytic leukemia cells. Oncotarget. 2016;**7**(29):46557-46572

[84] Pannkuk EL et al. A metabolomic serum signature from nonhuman primates treated with a radiation countermeasure, gamma-tocotrienol, and exposed to ionizing radiation. Health Physics. 2018;**115**(1):3-11

[85] Nie JH et al. Uncovering the anti-proliferation mechanism and bioactive compounds in red kidney bean coat against B16-F10 melanoma cells by metabolomics and network pharmacology analysis. Food & Function. 2019;**10**(2):912-924

[86] Bi W et al. Chemopreventive effects of Ku-jin tea against AOM-induced precancerous colorectal lesions in rats and metabolomic analysis. Scientific Reports. 2017;**7**(1):15893

[87] Lall RK et al. Dietary flavonoid fisetin increases abundance of high-molecular-mass hyaluronan conferring resistance to prostate oncogenesis. Carcinogenesis. 2016;**37**(9):918-928

[88] Yang CC et al. Plant galactolipid dLGG suppresses lung metastasis of melanoma through deregulating TNF-alpha-mediated pulmonary vascular permeability and circulating oxylipin dynamics in mice. International Journal of Cancer. 2018;**143**(12):3248-3261

[89] Chao WW et al. Phyto-sesquiterpene lactone deoxyelephantopin and cisplatin synergistically suppress lung metastasis of B16 melanoma in mice with reduced nephrotoxicity. Phytomedicine. 2018;**56**:194-206

[90] Liu R et al. Plasma N-acetylputrescine, cadaverine and 1,3-diaminopropane: potential biomarkers of lung cancer used to evaluate the efficacy of anticancer drugs. Oncotarget. 2017;**8**(51):88575-88585

[91] Liu R et al. Quantitative metabolomics for investigating the value of polyamines in the early diagnosis and therapy of colorectal cancer. Oncotarget. 2018;**9**(4):4583-4592

[92] Katsila T, Patrinos GP. Editorial: (Pharmaco)Metabolomics in drug discovery and individualisation of treatment. Current Pharmaceutical Design. 2017;**23**(14):2027

[93] Nicholson JK, Wilson ID, Lindon JC. Pharmacometabonomics as an effector for personalized medicine. Pharmacogenomics. 2011;**12**(1):103-111